AF454169

ALIENATION AMONG STUDENTS

ABOUT THE BOOK

This book is the outcome of the research work carried to know how school and home environment leads to alienation among school going students. Adolescence is a stage in the life span through which individuals pass in their preparation for adulthood. It is especially dynamic period because many of the roles of adolescents learning are unique to this time of life. This book is an attempt to find out the factors of home and school environment which leads to alienation. The book also brings out the influence of schools among the alienated students. This study is of paramount importance as it also tries to find out the reasons of drop outs among students. The research was funded by ICSSR, Government of India.

ABOUT THE AUTHOR

Dr. Fazli Muneer Rizvi completed her education from prestigious Aligarh Muslim University. She did her post doctoral fellowship from ICSSR, Govt. of India. She is the first post doctoral fellow from the Department of Psychology, Aligarh Muslim University. She has got more than seven years of teaching and research experience. She has published many papers in National and International journals. She takes keen interest in subjects of Child Psychology and Human Resource Management. Presently she is based in Saudi Arabia.

ALIENATION AMONG STUDENTS

By
DR. FAZLI MUNEER RIZVI

2015
Scholars World

A Division of

Astral International Pvt. Ltd.
New Delhi – 110 002

Published by : **Scholars World**
A Division of
Astral International Pvt. Ltd.
– ISO 9001:2008 Certified Company –
4760-61/23, Ansari Road, Darya Ganj
New Delhi-110 002
Ph. 011-43549197, 23278134
E-mail: info@astralint.com
Website: www.astralint.com

Laser Typesetting : **SSMG Computer Graphics**
Delhi - 110 084

Printed at : **Replika Press Pvt. Ltd.**

PRINTED IN INDIA

Dedicated to my Loving Parents

PREFACE

First of all, I am grateful with my heart and soul to the **"ALMIGHTY"** the most Beneficent and the most Merciful, who with His miraculous and bountiful blessings enabled the trivial existence of mine to do this job.

The present book, *"Alienation among Students"* is a result of my post doctoral research work where I examined *Alienation Among Students.* I have seen many students year after year dropping out of schools and discontinuing their. education it always troubled me and this was the reason I took this research.

Adolescence is an important phase in the life cycle. The development of familial, interpersonal, and institutional relationships at this critical stage in life may have lasting influences throughout the life-course. Adolescence is popularly described as a time of heightened egocentrism, volatility and experimentation with risky behaviors. Close emotional ties to parents are challenged, as adolescents begin to exercise their independence and individuality. By their very nature, adolescents involve many physiological, psychological, social and cognitive changes. These changes include formation of personal identity, the establishment of new networks, and a general orientation toward self and world.

Adolescence is a stage in the life span through which individuals pass in their preparation for adulthood. It is especially dynamic period because many of the roles of adolescents learning are unique to this time of life. The capacity to achieve a high level of self awareness arises during adolescence and self examination is the key to attaining maturity. Energies must be marshaled carefully. However, because consequences are cumulative, options are finite and every choice reduces later freedom.

While all adolescents experience potentially alienating contexts, there appears to be a threshold at which adolescents are no longer able to tolerate what is happening to them, and they exhibit behaviors commonly associated with alienation (Tripp, 1986).

I would like to describe briefly the main variables used in the present study such as alienation, home environment and school environment.

Alienation has been used in a variety of ways by philosophers, theologian, sociologist and psychologists. The alienated person as described by Fromm (1963), does not experience himself as the centre of his world, as the creator of his own acts but his acts and their consequences have become his masters, whom he obeys, or whom he even worship. For Seeman (1959) alienation has five components, namely, powerlessness, meaninglessness, normlessness, isolation and self estrangement, later Seeman (1972) revised alienation in six categories: (a) powerlessness – the sense of low control Vs. mastery over events, (b) meaninglessness the sense of incomprehensibility Vs. understandings of personal and social affairs, (c) normlessness high expectancies for (or commitment to) socially unapproved means Vs. conventional means for the achievement of given goals, (d) cultural estrangement called value isolation the individual's rejection of commonly held value is the society Vs. commitment, (e) self estrangement the individual's engagement in activities that are not intrinsically rewarding Vs. involvement in a task of activity for its own sake, and (f) social isolation – the sense of exclusion or rejection Vs. social acceptance (Seeman, 1975).

Alienation is a term used to describe student estrangement in the learning process (Brown, Higgins, & Paulsen, 2003). Mann (2001) defined alienation as "the state or experience of being isolated from a group or an activity to which one should belong or in which one should be involved". The alienated person as described by Fromm (1963), does not experience himself as the centre of his world, as the creator of his own acts but his acts and their consequences have become his masters, whom he obeys, or whom he even worship. Alienation can affect adolescents in their home, social life, and future opportunities however, this study was most concerned with non-completers' experiences of alienation from a high school setting (Bronfenbrenner, 1986).

Adolescents spend a large proportion of their day in school or pursuing school-related activities. While the primary purpose of school is the academic development of students, its effects on adolescents are far broader, also encompassing their physical and mental health, safety, civic engagement, and social development. Further, its effects on all these outcomes are produced through a variety of activities including formal pedagogy, after-school programs, caretaking activities (e.g., feeding, providing a safe environment) as well as the informal social environment created by students and staff on a daily basis. While most reports focus on a particular aspect of the school environment (e.g., academics, safety, health promotion), this brief looks at schools more comprehensively as an environment affecting multiple aspects of adolescent development.

A safe environment is a prerequisite for effective learning, so much so that the country's major education reform initiative, No Child Left Behind, requires school systems to have programs in place to reduce levels of violence as part of its larger plan to improve academic performance (U.S. Department of Education, 2007).

When students find their school environment to be supportive and caring, they are less likely to become involved in substance abuse, violence, and other problem behaviors (Hawkins, Catalano, Kosterman, Abbott, & Hill 1999; Battistich & Hom 1997; Resnick et al. 1997). They are more likely to develop positive attitudes toward themselves and prosocial attitudes and behaviors toward others (Schaps, Battistich, & Solomon 1997). Much of the available research shows that supportive schools foster these positive outcomes by promoting students' sense of "connectedness" (Resnick et al. 1997), "belongingness" (Baumeister & Leary 1995), or "community" (Schaps, Battistich, & Solomon 1997) during the school day.

The role of home environment in fostering positive academic beliefs (values, competence perceptions, and goals) and behaviors is seen as important to keeping students engaged in school during adolescence. Research has shown that parents' beliefs and values can have an important impact of their children's achievement-related beliefs and values (Ames

& Archer, 1987; Parson, Adler & Kaczala, 1982). In addition, direct parent involvement in and support of the child's school experiences are known to play an important role in a child's academic success (Epstein, 1989).

When a child stays at home all day, and parents take on the dual role of teacher and parent, issues of discipline will arise. It is easy for the child to take on a negative attitude towards understanding discipline. Correct discipline needs to be adhered from the start of homeschooling, to avoid potential difficulties later on in the child's development and learning.

The research was systematically designed in accordance with the following main research objectives:

1. To identify alienation among public school students.

2. To identify alienation among convent school students.

3. To examine the difference between mean scores of Public and Convent School students on alienation.

4. To examine the difference between mean scores of Public and Convent School students on School Environment.

5. To examine the difference between mean scores of Public and Convent School students on Home Environment.

6. To identify the influences of public school and convent school students on alienated adolescents.

The data sample for this study consists of 300 adult students. Subjects were drawn from Al-Barkaat Public School and convent; Our Lady of Fatima School of Aligarh (UP), India. Subjects were classified on the basis of scores obtained on alienation scale on which Quartile 1 and Quartile 3 were calculated to get the alienated subjects. Public school subjects whose scores fell below Q1 (45) were identified as low alienated and subjects whose scores fell above Q3 (52) were categorized as high alienated. Convent school subjects who have obtained scores below 41 were categorized as low alienated and those subjects whose score were above 49 they categorized as high alienated subjects. The sample further classified into two categories i.e. 50 students were drawn from convent schools and other 50 were taken from public schools.

The following tools were used in the present study

Alienation scale developed by Kureshi and Dutt (1979) was used in this study..This scale comprised 21 items with four alternative response categories (always, often, sometimes, never, representing the five factors despair, disillusionment, unstructured universe, psychological vacuum and narcissism.

The Home and School Environment questionnaires are subscales of The Resilience and Youth Development Module (RYDM) which is a component of California Healthy Kids Survey (WestEd, 2002). The RYDM is devoted completely to assessing the internal and external assets associated with positive youth development and resilience (WestEd, 2002). The RYDM provides comprehensive and balanced coverage of external assets in home and school environment.

The full RYDM contains 59 questions that measure 17 external and 6 internal assets in the home, school, community, peer group and in the individual domains. Both, the home and school environment scales for measuring external assets, have 9 items each, and measure three common dimensions: *Caring relationships, High expectations and Meaningful participation.* Each item has four response options (very much true, pretty much true, a little true and not at all true) out of which the participants had to choose the one option which they felt best applied to them.

I established rapport with the respondent students, prior to administer the proposed scales, and assured them that their responses would be kept strictly confidential and would be utilized for the research purpose only. After establishing rapport with the students, the data were collected in their classrooms in many sessions.

The main findings of this study are as follows:

- 'Expectation in home'- as a dimension of home environment was found significant predictor of alienation among students.

- 'Caring relationship in school' was negatively related to alienation whereas 'Total home environment positively related to alienation in public school students.

- 'Meaningful participation in home' was found negatively related with alienation in Convent school students.

- Public school students differ significantly with Convent school students on 'caring relationship in school', 'meaningful participation in school' –as a dimensions of school environment, 'home environment as a whole' and it's all three dimensions namely 'caring relationship in home', 'expectation in home', 'and meaningful participation in home'.

- Both school students (Public and Convent) did not differ significantly with each other on 'Alienation', and also did not differ with regards to 'School environment as a whole' and 'Expectation in school'.

- Male students differed significantly with female students on 'caring relationship in school', 'home environment as a whole' and its two dimensions namely, 'expectation in home', 'and meaningful participation in home'.

- Both groups (male and female) did not differ significantly on 'Alienation', 'School environment', 'Expectation in school', 'Meaningful participation in school' and 'Caring relationship in home'.

I hope this book will enhance the understanding of alienation among school going children and help further in reducing the drop outs and decline in their performance

Happy reading

Al- baha, Saudi Arabia   Dr Fazli Muneer Rizvi

ACKNOWLEDGEMENTS

"True knowledge exists in knowing that you know nothing".

I would like to express my gratitude to all the people who saw me through this book; to all those who provided support, talked things over, read, wrote, offered comments, and assisted in the editing, proofreading and design.

I would like to thank Dr. Mohd. Ilyas Khan & Prof. Akbar Husain for their expert advice encouragement, keen interest and thought provoking guidance throughout this difficult project

This project would have been impossible without the financial support of Indian Council of Social Science Research. Government of India

To thanks parents have no meaning because my firm belief is to accomplish a task successfully is in itself gratitude towards parents. At this juncture I remember my Ammi, my brother, my husband Dr. Mohammed Abul Khair and loving in-laws.

Last and not least: I beg forgiveness of all those who have been with me over the course of the years and whose names I have failed to mention."

Dr. Fazli Muneer Rizvi

CONTENTS

List of Tables

Chapter 1
INTRODUCTION

1.1 Background of the study

Throughout the 20th century, the concept of alienation has received substantial attention in the social sciences. Alienation research peaked during the 1970s and has received declining attention until recently. This concept has been criticized as being used too broadly to describe nearly any kind of aberrant behavior, "ranging from political manifestations to psychopathology" Lacourse et al., (2003). Alienation may be viewed as a result of pervasive social forces beyond school, such as specialization, mobility, bureaucratization, capitalism, or other features of the modern world that fragment human experience (Davison Aviles, Guerrero, Howarth, & Thomas 1999). Alienation is a term used to describe student estrangement in the learning process (Brown, Higgins, & Paulsen, 2003). Mann (2001) defined alienation as "the state or experience of being isolated from a group or an activity to which one should belong or in which one should be involved".

According to Dean (1961) alienation can be influenced by a variety of factors, including factors from the home and from within the adolescents themselves. Alienation among adolescents has been primarily attributed to school organization and the adolescent's home environment (Bronfenbrenner, 1974, 1986; Rafalides & Hoy, 1971).

Several studies have showed that poor -teacher relationships are a major cause of student's alienation from school, which in turn may lead to dropping out of high school (Jordan, Lara, & McPartland 1996). School becomes a toxic environment for many students because it represents a constant source "of failure and frustration at a time of development when they need a sense of success and a positive image of themselves in relation to a complex world" (Wehlage, 1986).

Students may select such behaviors based on the discord that may exist in their personal worlds, however, our general understanding of the personal experiences of alienated students such as Kip Kinkel, the young man who in 1998 killed both his parents and then shot and killed two students and injured 25 others at Thurston High School in Springfield, Oregon, indicated that schools did play a role in exacerbating the individual's sense of alienation (Edwards & Mullis, 2001; Smokowski & Kapasz, 2005). Both teachers and administrators contribute to the creation and maintenance of a school culture which may undermine the efforts of less privileged and already reluctant learners (Wehlage, 1986).

In exploring how students' family background may influence the formation of supportive student-teacher relationships at school, many studies have pointed to the importance of students' family cultural environment. These studies have focused on how family cultural environment may influence the standards that educators use to evaluate students and their parents (Kingston 2001; Lareau & Weininger 2003; Reay 2004).

The present study would focus on what the role of psychologist that can develop better understanding of alienated students. Indeed, alienated students have been affected by inability to cope with the school environment and also home environment so, school environment and home environment have been found as contributing factors to the student's life.

In recent days, student's alienation is a major problem. In the proposed study, researcher wishes to explore the relationship of school environment and the home environment with sense of alienation and its impact on the mind of the students. Home school relation is the new field, and very little research has been done so far. Schools play vital roles in the child's overall personality development. For instance, learning develops, how to control emotions, cope with different situations, recognizing their positions, status and the development of values and their responsibilities to society. In this proposal, "Term Alienation" is used in the context of school going children; therefore, we are referring to the phenomenon as 'Student's Alienation'.

Adolescence marks the transition between childhood and adulthood. By its very nature, it involves many physiological, psychological, social, and cognitive changes. These changes include the formation of a personal identity, the establishment of new peer networks, and the development of abstract thinking skills (Dacey & Kenny, 1997; Geldard & Geldard, 1999; Heaven, 1994). To manage these challenges, adolescents rely on their coping repertoire, which includes their problem solving competencies and skills.

Adolescent is at best a difficult, awkward period (Calabrese, 1987; Gulltta, 2001). It's a time when children move toward adulthood in tentative and often faltering steps. It can be considered "a process of individuation and identity creation that necessarily implies discomfort, discontent, and bouts of real and imagined alienation and ostracism" (Sikkink, 1999). It is inability of adolescents to connect meaningfully with other peoples. Brown, et al. (2003) defines adolescent alienation "as a feeling of aloneness, a feeling that no one likes them, and that they are not what others want them to be".

Many variables can trigger an adolescent's feelings of isolation and disconnection such as changes in family circumstances, lack of social acceptance, and academic underachievement (Hawkins et al. 2000). Alienation can affect adolescents in their home, social life, and future opportunities (Bronfenbrenner, 1986); however, this study was most concerned with non-completers' experiences of alienation from a high school setting.

The present study is aimed at finding out the "Alienation of students in Public and Convent school students in relation to School Environment and Home Environment". The variables like alienation, home environment and school environment need to be describing briefly. Before going into detail with regards to the variables used in the study, it will be fitness of things to describe the word 'Adolescence' that is the sample area of the present study.

1.2. Adolescence

"Adolescence" come from the Latin term "adolescre" which means to "Grow or to Grow to maturity." Broadly speaking during adolescence there are individual differences and it is possible to mark to mark off period within adolescent span.

Adolescence is a stage in the life span through which individuals pass in their preparation for adulthood. It is especially dynamic period because many of the roles of adolescents learning are unique to this time of life. The capacity to achieve a high level of self awareness arises during adolescence and self examination is the key to attaining maturity. Energies must be marshaled carefully. However, because consequences are cumulative, options are finite and every choice reduces later freedom.

A comprehensive understanding of adolescence also requires knowledge of youth's consciousness or the personal factors, aspirations, attitudes, beliefs, dispositions that enable young people to sustain intimate and lasting social relationship, accept vocational and economic responsibilities and form ideological convictions.

All young people have basic needs that are critical to survival and healthy development. They include a sense of safety and structure, belonging and membership; self worth and an ability to contribute; independence and control over one's life; closeness and several good relationships and competency and mastery. At the same time to succeed as adults all youth must acquire positive attitudes and appropriate behavior and skills in five areas: health, personal/social knowledge, reasoning and creative, vocation and citizenship. (Politz, 1996).

Hoffman, Levy, & Malinsky, (1996) indicate that the problem arises due to the transitional aspect during the adolescent period and as a bridge between childhood and adulthood. From the beginning of adolescence, the average youth is preoccupied with problems related to vocation, parental encouragement adjustment in the major areas of adult life.

There are many reasons why the problems of adolescence are so difficult that are especially common.

- ◆ Very few young people have had any preparation for meeting the type of problems they are expected to cope as adults. Education in high school and college provides only limited knowledge for jobs, and few schools or colleges give courses in the common problems of vocation, family and parenthood. ?

♦ Just as trying to learn two or more skills simultaneously usually results in not learning any one of them well, so trying to adjust to two or more new roles simultaneously results in poor adjustment to all of them. It is difficult for the adolescent to deal with a choice with a career.

♦ Young people do not that they had when they were younger. This is partly their fault and partly the fault of their parents and teachers.

♦ Most of the adolescents are too proud of their own new status to admit that they cannot cope with it. So they do not take the advice and help in meeting the problems this new status give rise to. Similarly, most parents and teachers, having being rebuffed by adolescents who claimed capable of handling their own affairs hesitate to offer help unless they are specially asked to do so. The shortening of adolescence has made the transition to adulthood especially difficult.

Adolescent is the essence and spring of life. It is the period when an individual is on the threshold to adulthood. Earlier civilization did not consider puberty and adolescent to be distinct period in life span. A child tern adolescent has a broader meaning, and mincludes mental, emotional, social and physical maturity.

Adolescence was regarded as beginning when the individual becomes sexually mature and ending when he reaches legal maturity. However, studies of changes in behavior through-out adolescence have revealed not only that these changes are more rapid in the early than in the later part of adolescence but also that behavior and attitudes in the early part of the period are markedly different from those in the later part. As a result, it has become a widespread practice to divide adolescence into two period i.e. early adolescence and late adolescence. Early adolescence extends roughly from thirteen (13) to sixteen (16) or seventeen (17) years and late adolescence cover the period from seventeen (17) until eighteen (18), the age of legal maturity.

Some individuals and some people mature early while others enjoy, or suffer a considerable period of growth after childhood before, since the full development of power and abilities has been achieved before life settles down into routine of an adjusted efficient, adult age, is the period which as some to be termed "adolescence." It is one of the normal stages of developments; certain marked changes in different areas of developments put forth extra demands on the adolescents for proper adjustment. Further, the complexity and the changes of the modern society expect the adolescent to interact suitable and efficiently to cope up with challenging situations.

1.3. Alienation

Alienation has been used in a variety of ways by philosophers, theologian, sociologist and psychologists. The alienated person as described by Fromm (1963), does not experience himself as the centre of his world, as the creator of his own acts but his acts and their consequences have become his masters, whom he obeys, or whom he even worship. For Seeman (1959) alienation has five components, namely, powerlessness, meaninglessness, normlessness, isolation and self estrangement, later Seeman (1972) revised alienation in six categories: (a) powerlessness – the sense of low control Vs. mastery over events, (b) meaninglessness the sense of incomprehensibility Vs. understandings of personal and social affairs, (c) normlessness high expectancies for (or commitment to) socially unapproved means Vs. conventional means for the achievement of given goals, (d) cultural estrangement called value isolation the individual's rejection of commonly held value is the society Vs. commitment, (e) self estrangement the individual's engagement in activities that are not intrinsically rewarding Vs. involvement in a task of activity for its own sake, and (f) social isolation – the sense of exclusion or rejection Vs. social acceptance (Seeman, 1975).

The foregoing theoretical analysis of the conceptual basis of this investigation leads us to believe that we need a reinculcation of certain values, which, apart from having other imperatives, include non-violence, love, and mutual acceptance. Only then perhaps we can hope to alleviate the present situation under which are inching into a social and cultural death.

Alienation may be viewed as a result of pervasive social forces beyond school, such as specialization, mobility, bureaucratization, capitalism, or other features of the modern world that fragment human experience

(Davison, Aviles, Guerrero, Howarth, & Thomas, 1999). Alienation is a term used to describe student estrangement in the learning process (Brown, Higgins, & Paulsen, 2003). Mann (2001) define abstract alienation as "the state or experience of being isolated from a group or an activity to which one should belong or in which one should be involved.

Normlessness reflects lack of appropriate rule-governed behavior (e.g., academic dishonesty). Meaninglessness describes alienated students' interpretation of curriculum as irrelevant to their current and future needs. Loneliness and separation from peers and teachers characterizes social isolation. Alienation is a useful construct for understanding the mechanisms associated with undesirable learner outcomes and in developing strategies to circumvent student academic failure (Taylor, 1999).

Langman and Kalekin-Fishman (2006) viewed social alienation as rooted in dysfunctional communication and power within a civil society. Alienation was returned to an individual need and social response for recognition. While the dimensions of alienation varied depending upon age, socio-economic status, race, and gender, an underlying commonality existed, the relationship between the individual and society.

The sociological issue of alienation was defined within a psychological paradigm and suggested a mind-body relationship to one's social setting. Extending the soul into the paradigm was problematic, as the perception of the mind was often linked to the human lifecycle, whereas the soul was associated less with the cognitive functions and more with the spiritual realm (Richert & Harris, 2008). Nonetheless, there has been a recent upsurge of interest in applying the concept to adolescents' experiences, most specifically their scholastic experience (Alienation in school settings can best be described as a process (Carlson, 1995) or contextually relevant rather than a personality trait or a state of being.

While all adolescents experience potentially alienating contexts, there appears to be a threshold at which adolescents are no longer able to tolerate what is happening to them, and they exhibit behaviors commonly associated with alienation (Tripp, 1986). Bronfenbrenner (1986) described alienation as a lack of a sense of belonging, and feeling cut off from family, friends, or school; it is the inability to connect meaningfully with other people. For adolescents, "alienation signifies a separation or distance among two or more entities and involves a sense of anguish or loss, resulting in a student viewing life and school as fragmentary and incomplete" (Brown et al., 2003). And while many discussions in the literature explain various strategies for decreasing adolescent alienation in schools (Brown et al., 2003), there is little discussion regarding how the alienated students themselves experience and make meaning of the process of becoming and being alienated. Mau (1992) identified four dimensions that have guided the research of alienation in schools. They are:

(a) powerlessness, the inability to control the forces that surround one's life,

(b) meaninglessness, the inability to see purpose in one's life or work,

(c) normlessness, the refusal to accept life's societal restrictions, and

(d) social estrangement, the separation from significant others. All of these dimensions relate to common problems in school such as absenteeism, low commitment to school work, and academic underachievement (Calabrese, & Adams 1990).

Powerlessness manifests when a student places value on a goal, but at the same time has low expectations of realizing that goal (Mau, 1992). Teachers and administrators have all the power to indicate whether or not a student is successful, and students become increasingly more critical and less tolerant of this authority (Rafalides & Hoy, 1971).

According to Brown et al. (2003) "when students believe that, they can personally do to influence their future in school, they disengage from the process." These students learn that their complaints and concerns go unheard and eventually rebellious behaviors such as skipping class, not doing or turning in homework, and thwarting school norms and rules replace efforts to learn (Mau, 1992).

Meaninglessness refers to the issue of relevancy and significance. Students may perceive a limited relationship between the present expectations of school work and the future requirements of job and adulthood (Mau, 1992). One can look to the continuing conflicts arising from Oregon's School to Work movement as well as the No Child Left Behind legislation to understand how meaningful working for an intangible future outcome (or a CIM certificate) is to modern students, and especially those with already limited aspirations. Students, who do not feel important or significant in positive ways, will often look for importance in negative (Edwards & Mullis, 2001).

Normlessness refers to the belief on the part of students that socially disapproved of behavior is necessary in order to achieve school goals (Mau, 1992). Brown et al. (2003) provide examples such as cheating on a test in order to get a good grade, praise from an adult, or believing that a D is an acceptable grade. The insight that Mau offers in describing this behavior is as follows: Organizational practices of meritorious grading and curriculum grouping (tracking) as well as teacher expectations not only create different learning environments, but accentuate academic differences among students (McPartland & McDill, 1982). Through social comparison processes, students in marginal positions are not integrated into the normative school structure.

Social estrangement is the result of a student's inability or unwillingness to integrate into the school's social existing network (Brown et al., 2003). A student who feels estranged typically doesn't accept the goals and values of a school and tends to reject school and all that it stands for (Rafalides & Hoy, 1971). Social estrangement is the result of a student's inability or unwillingness to integrate into the school's social existing network (Brown et al., 2003). A student who feels estranged typically doesn't accept the goals and values of a school and tends to reject school and all that it stands for (Rafalides & Hoy, 1971).

1.4. Alienation in Adolescent

Adolescence is at best a difficult, awkward period (Calabrese, 1987; Gullotta, 2001). It's a time when children move toward adulthood in tentative and often faltering steps. It can be considered "a process of individuation and identity creation that necessarily implies discomfort, discontent, and bouts of real and imagined alienation and ostracism" (Sikkink, 1999). The tumultuous process of being an adolescent coincides with the feelings often associated with the process of) becoming alienated. Alienation can affect adolescents in their home, social life, and future opportunities however, this study was most concerned with non-completers' experiences of alienation from a high school setting (Bronfenbrenner, 1986).

1.4.1. Student's Alienation

Alienated students are described as those individuals who live on the borders of the school (Smith & Goc-Karp, 1994), the student who considers him or herself an outcast socially (Rodriguez, 1997) and academically, or "students who are reluctant to maintain the charade of acceptable behavior in school" (Carley, 1994). An alienating school environment can produce negative and disruptive behaviors which may lead to non-completion. These behaviors include hostility, passivity, withdrawal, lack of sense of responsibility for learning and getting work done, poor quality work, disinterest, and lack of involvement and initiative.

Williamson and Cullingford, (1997) and Gullotta (2001) suggests certain stresses of adolescence such as hormonal changes, preoccupation with self, and peer relationships in combination with a school system which tends toward discriminatory, sexist, and elitist practices may contribute to the alienation of students.

1.4.2. Who are the Alienated

When discussing the construct of alienation, it would be remiss not be appropriate to discuss who are alienated students are? Rodriguez (1997) wrote that historically students at risk of school failure are those "whose appearance, language, culture, values, communities, and family structures do not correspond to those of the dominant white culture that schools were designed to serve and support". Not all students claiming a sense of alienation in schools are students of color or of poverty; however, in the case of student alienation these populations tend to be overrepresented (Goodwin, 2000; Kawakami, 1994; Rodriguez, 1997). Viewing non-completion as a possible resultant behavior of school alienation may explain the high number of underrepresented youth who express feelings of non-attachment and lack of relevancy in public high school (Kawakami, 1994).

1.5. School Environment

Adolescents spend a large proportion of their day in school or pursuing school-related activities. While the primary purpose of school is the academic development of students, its effects on adolescents are far broader, also encompassing their physical and mental health, safety, civic engagement, and social development. Further, its effects on all these outcomes are produced through a variety of activities including formal pedagogy, after-school programs, caretaking activities (e.g., feeding, providing a safe environment) as well as the informal social environment created by students and staff on a daily basis. While most reports focus on a particular aspect of the school environment (e.g., academics, safety, health promotion), this brief looks at schools more comprehensively as an environment affecting multiple aspects of adolescent development. Research has repeatedly demonstrated the interconnectedness of the pieces, with safety and health affecting the academic environment, academics affecting health and social development, and so on (Scott, et al. 2001; Geierstanger, et al. 2004; Roeser & Eccles, 1998). For that reason, any particular aspect of school policy and activities will be better understood through the lens of that larger context. This is particularly important as school systems have become even more pressured to focus on their main goal of academic development as a result of the federal No Child Left Behind initiative.

A safe environment is a prerequisite for effective learning, so much so that the country's major education reform initiative, No Child Left Behind, requires school systems to have programs in place to reduce levels of violence as part of its larger plan to improve academic performance (U.S. Department of Education, 2007).

While schools are called on to shape many aspects of students' lives, their core focus is clearly the development of academic knowledge and skills. A recent review of the literature recommended five key indicators of school environmental quality because of their strong links to student learning (Mayer & Ralph, 2007). These include: teacher's academic skills; teacher experience; demanding course content; access to technology (especially computers and the Internet); and class size. Teachers are more effective and student outcomes are better when teachers have more experience, teach in the field in which they are trained, and when they have strong academic skills. There is some evidence that in small classrooms teachers is able to provide more opportunity for participation, spend more time on instruction, and is faced with fewer disciplinary problems (U.S. Department of Education, 2000). Regarding the relationship between class size and student achievement, several Meta analyses have concluded that students in smaller classes have higher test scores and that the effect may be even larger for disadvantaged students.

Researchers have consistently cited the school environment as a major cause of adolescent's alienation (Calabrese, 1987) and have identified factors which contribute to the level of alienation experienced by students (Brown et al., 2003). Some of the school factors that have been linked with increased student's alienation are:

1. A student's perception of a lack of control; a school's maintenance of high control environments (Calabrese & Seldin, 1986);

2. The degree of social acceptance (Taylor, 1999) and peer to peer relationships (Valverde, 1987);

3. Teacher expectations and value placed on school work (Kunkle, Thompson, & McElhinney, 1973);

4. The perceived relevancy of the school curriculum (Kunkel et al, 1973);

5. School culture (Calabrese & Seldin, 1987; Calabrese & Noboa, 1995); and

6. Discriminatory practices based on race, ethnicity, linguistic difference, socioeconomic status, ableness, intelligence, and gender (Brown et al., 2003).

Feeling alienated from the school environment is a negative consequence of these factors or combination of factors which exists in schools (Calabrese & Noboa, 1995; Staples, 2000; Wehlage & Rutter, 1986) and will continue to "negatively affect the   self-confidence, self-esteem, and self-direction of some students" (Brown et al., 2003). Behaviors which may indicate a student might feel alienated from the system are numerous and varied. McMillan (1992) suggested the following list:

a) Behavior problems – "in trouble" in school or community, acting out behavior, disruptive in learning environment.

b) Potentiality to attempt or complete suicide.

c) Absenteeism.

d) Lack of respect for authority, feelings of alienation from   school authorities.

e) Grade retention – especially in early grades.

f) Suspensions/expulsions.

g) Course failure, poor academic record.

h) Tracking/ability grouping.

i) Dissatisfaction and frustration with school.

j) Lack of available and adequate counseling opportunities.

k) Inadequate school services – mental health, social services and health services,

l) School climate hostile to students who do not "fit the norm"

School becomes a toxic environment for many students because it represents a constant source "of failure and frustration at a time of development when they need a sense of success and a positive image of themselves in relation to a complex world" (Wehlage, 1986). Using the High School and Beyond dataset, Wehlage identified three variables which indicate how schools send out signals to youth that they are "neither able nor worthy enough to continue to graduation". The three indicators are as follows: (a) perceived teacher interest in students, (b) the effectiveness of the school's discipline system, and (c) the fairness of the school's discipline system. In other words, alienated students, even decades after Wehlage's statement, continue to experience the realities of being powerless, perceiving no relevant meaning to class and school activities, not buying into authoritative norms, and feeling socially isolated.

The focus on and value of these student experiences can contribute greatly to our general understanding of the relationship between students' behavior and a school's climate, peer to peer relations. For many young people, the school experience is filled with ambiguity. School is a place that can be incredibly alienating, but at the same time, it can be an important place for social connection with peers, teachers, and administrators (Brown et al., 2003).

While the main purpose of education is to prepare adolescents academically, schools are increasingly called upon to develop socially competent, physically healthy and civically engaged youth who will also carry those assets into adulthood. The school provides many adolescents with a common meeting place, and often times a refuge from family or community tension and potential harm. Nevertheless, school as a socializing place can be a negative experience for many indicating that something about school and how it functions is wrong (Calabrese, 1986).

The need to belong is powerful and when feeling rejected an adolescent is not likely to respond well to adult intervention, either at school or at home (Smokowski & Kapasz, 2005). In a hostile or toxic school culture, many students become victims of peer to peer abuse such as harassment, bullying, and relational aggression. Left unchecked or unobserved by adults, these students become isolated, untrusting, marginalized, and alienated (Smokowski & Kapasz, 2005).

A substantial body of researches show that, for good or ill, a school's social environment has broad influence on students' learning and growth, including major aspects of their social, emotional, and ethical development. The social environment is shaped by many factors:

- The school's espoused goals and values
- The principal's leadership style
- The faculty's teaching and discipline methods
- The policies regarding grading and tracking
- The inclusion or exclusion of students and parents in the planning and
- decision-making processes

But perhaps most important in determining the school environment is the quality of students' relationships with other students and with the school's staff.

The importance of the school environment is underscored by the Search Institute's list of environmental and individual "developmental assets" that serve as general protective factors (Leffert, Benson, & Roehlkepartain 1997). Among the items in the institute's list of environmental assets are:

- A caring school climate
- Parental involvement in schooling
- Clear rules and consequences in the school and family
- High expectations from teachers and parents

Among the items in the institute's list of individual assets are:

- Motivation to achieve
- School engagement
- Bonding to school

When students find their school environment to be supportive and caring, they are less likely to become involved in substance abuse, violence, and other problem behaviors (Hawkins, Catalano, Kosterman, Abbott, & Hill 1999; Battistich & Hom 1997; Resnick et al. 1997). They are more likely to develop positive attitudes toward themselves and prosocial attitudes and behaviors toward others (Schaps, Battistich, & Solomon 1997). Much of the available research shows that supportive schools foster these positive outcomes by promoting students' sense of "connectedness" (Resnick et al. 1997), "belongingness" (Baumeister & Leary 1995), or "community" (Schaps, Battistich, & Solomon 1997) during the school day.

1.5.1. School Climate

A large body of literature is dedicated to the examination of the factors that contribute to academic failure. Of growing interest to that discussion is the role school climate plays in student behaviors and attitudes. School climate can be defined as "the collective personality of a school; the atmosphere as characterized by the social and professional interactions of the individuals in the school" [American School Counselor Association (ASCA), 2003].

Researchers have used various definitions of school climate. Hoy and Miskel (2005) define school climate as "the set of internal characteristics that distinguish one school from another and influence the behaviors of each school's members". A positive school climate can enhance staff performance, promote higher morale, and improve student achievement (Frieberg, 1998; Kelley, Thornton, & Daugherty, 2005) and studies by Heck (2000) and Goddard, Hoy and Hoy (2000) link school climate and student achievement (Kelley et al., 2005).

Current literature on school climate seems to agree with the 1985 statement from Hoyle, English, and Steffy school climate may be one of the most important ingredients of a successful instructional program. Without a climate that creates a harmonious and well functioning school, a high degree of academic achievement is difficult, if not downright impossible to obtain.

Bulach, Malone, and Castleman (1995) found that a significant relationship exists between student achievement and school climate and Kelley et al. (2005) concluded that school climate is a significant factor in successful school reform. Academic achievement is related to schools with a strong academic emphasis within the context of open and healthy climates (Hoy, Tarter & Bliss, 1990) and "unless students experience a positive and supportive climate, some may never achieve that most minimum standards or realize their full potential" (Urban, 1999).

The overall school climate appears to have significant impact on a students' perception of their sense of belonging, contributing to either a feeling of "place attachment" (Swaminathan, 2004) or one of alienation. Integral to the creation of a welcoming and safe school climate, the classroom climate also has a powerful influence on student's connections to school.

1.5.2. The Relationship between School Climate and Student's Alienation

The literature supports the idea that the school environment and school processes impact the degree of student connectedness and alienation. And while much attention has been paid to the study of students' intellectual abilities and social development, "less attention has been paid to how well schools engage students in school life and how this affects their outlook on schooling and the future" (Thomson, 2005).

Currently, schools are deemed effective based on the percentage of students who achieve an acceptable score on state and federally mandated standardsbased assessments of reading and reading comprehension, writing, science, mathematics, and history. From a parent perspective, an effective school is one in which "students have learned the basic skills they need to enter college, get technical training, or enter the workforce" (Anderson & Cotton, 2001). In other words, if students aren't prepared for 'the next step", schools have failed. An effective school climate, however, is about more than maximizing academic achievement or university preparedness (Thomson, 2005).

According to Lashway (2003) the following three explanations have received the most attention:

(a). Demographics – Some schools serve predominately low income populations, incorporating an environment that may destabilize home lives, undermine support from home and community, and creates despair amongst the student body. In addition, "many low-income children are also members of racial minorities that face additional barriers to high achievement (Shannon & Bylsma, 2002).

(b). Insufficient Resources – Many states are suffering from an inability to provide stable funding for schools (Jerald, 2000). Lack of resources may account for destabilizing influences on students such as: low teacher morale, high classified and certified turnover, less alternative and after hours programming for failing students, and teachers forced to teach outside their specialties (Jerald, 2000).

(c). Ineffective school practices – Jerald (2000) identifies uncoordinated curriculum, superficial instructional strategies, scattershot professional development, and timid leadership as some of the factors that may hold school back from adequately supporting student academic and social growth.

In addition, lack of accountability by school officials is a contributing factor identified by Sullivan (2003) regarding poor quality of education and ineffective school climate. "Lack of accountability is closely tied to the school system's failure to ensure effective participation by parents and communities" (Sullivan, 2003) and by excluding parents and communities from participating, communities cannot hold schools accountable for the academic underachievement of its students.

Schools tend to operate based on the norms of the dominant, middle class whether they serve that population or not. Historically, public school administrators have not shared the same socioeconomic classes as their students (Alston, 2004), because a student's social class is strongly correlated with academic success (Cunningham & Cordeiro, 2000), the better understanding school leaders have of the life circumstances of the population it serves and the higher the involvement of the parents and community, the more likely students will feel connected and valued, and the higher they will achieve.

1.6. Student's Home Environment

A child's home cultural environment is defined as parents' willingness and capacity to help with their child' schooling. What is innovative in the present study is that it includes teachers' voices when examining teachers' evaluations and expectation of children, in addition to the conventional measures of family socioeconomic status teachers and counselors (Stanton-Salazar & Dornbusch, 1995).

The role of home environment in fostering positive academic beliefs (values, competence perceptions, and goals) and behaviors is seen as important to keeping students engaged in school during adolescence. Research has shown that parents' beliefs and values can have an important impact of their children's achievement-related beliefs and values (Ames & Archer, 1987; Parson, Adler & Kaczala, 1982). In addition, direct parent involvement in and support of the child's school experiences are known to play an important role in a child's academic success (Epstein, 1989).

Even though it is widely recognized in the educational community how important parent involvement is in helping schools to work effectively for all students (Comer, 1988), some evidence suggests that parents in this country actually feel that the school personnel, not themselves, are in large measure responsible for their child's education (Stevenson & Stigler, 1992). This attitude seems to become more prevalent as child progresses in school. Parent-school connections and parent involvement in general have been shown to decline right around the time students enter the middle or junior high school (Carnegie Council on Adolescent Development 1989; Epstein, 1983). This is especially problematic considering that the overlap between a home environment that promotes support for and valuing of education, and an educational setting that creates a sense of belonging, promotes competence, and instills autonomy are thought to be critical to the healthy development of adolescents' during this time (Bronfenbrenner, 1986; Eccles, Midgley, Wigfield, Buchanan, Reuman, Flanagan & MacIver, 1993).

When a child stays at home all day, and parents take on the dual role of teacher and parent, issues of discipline will arise. It is easy for the child to take on a negative attitude towards understanding discipline. Correct discipline needs to be adhered from the start of homeschooling, to avoid potential difficulties later on in the child's development and learning.

Discipline provides both parents and child with immense levels of freedom, and there will be an enticement to stretch this freedom. Certain rules and practices need to be implemented at the beginning stages of the child as it may be very difficult for parent to change their child's habits at later stage.

There should be a friendly and enabling environment at home. The members of the family should listen and have a great respect for one another. This is very important to maintain a good environment free from all misunderstandings and confusions among the family members. Parents' relations play a significant role in maintaining a better environment at home. Negative relations and disputes can spoil the atmosphere of a house. Instead of a better environment the entire house plunges into chaos which not only affects the psychology of children but also makes them suffer in various complexes.

We have stressed in this text that children are motivated to work on activities and learn new information and skills when their environments are rich in interesting activities that arouse their curiosity and offer moderate challenges. The same can be said about the home environment. Unfortunately, there is much variability in motivational influences in homes. Some homes have many activities that stimulate children's thinking, as well as computers, books, puzzles, and the like. Parents may be heavily invested in their children's cognitive development, and spend time with them on learning. Other homes do not have these resources and adults in the environment may pay little attention to children's education (Eccles et al., 1998).

There is much evidence supporting the present study that the quality of a child's early learning in the home environment relates positively to the development of intelligence and reading skills (Senechal & Lefevre, 2002), and parental involvement in schooling also predicts achievement (Englund, Luckner, Whaley, & Egeland, 2004). Various home factors have been shown to be important: mother's responsiveness, discipline style, and involvement with the child; organization of the environment; availability of appropriate learning materials; opportunities for daily stimulation.

Gottfried, Fleming, and Gottfried (1998) conducted longitudinal research examining the role of cognitive stimulation in the home on children's academic intrinsic motivation. Home environment variables measured included family discussions; attendance at cultural events; library visits; trips taken; importance of reading; provision of private lessons; access to play equipment; and family interest in music, art, and literature. The authors assessed home environment when children were age 8 and academic motivation at ages 9, 10, and 13. The results showed that children whose homes had greater cognitive stimulation displayed higher academic motivation from ages 9 through 13. The effect of SES was indirect: Families of higher SES were more likely to provide cognitively stimulating home environments, which in turn directly increased academic motivation. The fact that home environment effects were both short- and long-term suggests that home environment continues to play a role in early adolescence when peer influence becomes more powerful. These results highlight the need for parent awareness programs that teach them how to provide rich learning experiences for their children.

Within the home environment, we must examine both the roles of mothers and fathers, because differential parent behavior has often been implicated as a variable affecting children's development (Eccles et al., 1998). Eccles et al. (1998) listed six potential parental beliefs that can influence children's motivational beliefs:

1. attributions for the child's school performance,

2. perceptions of the task difficulty of schoolwork,

3. expectations and confidence in children's abilities,

4. values for schoolwork,

5. actual achievement standards, and

6. beliefs about barriers to success and strategies for overcoming these barriers.

Most of research on educational stratification in developing countries has focused on the impact of poverty, gender, and school quality on children's access to schooling and school outcomes (Adams & Hannum, 2007; Fuller & Clark, 1994; Fuller, 1987; Hanushek, 1995; Zhang, Kao, & Hannum, 2007). The teachers' evaluation of students' learning capacity and behavior may impact how teachers interact with students in the classroom (Hauser-Cram, Sirin, & Stipek, 2003). Meanwhile teachers' evaluations have strong influence on children, whether these evaluations are accurate or not (Entwisle, Alexander, & Olson, 1997; Downey & Pribesh, 2004; Hallinan, 2008; Hauser-Cram, Sirin, & Stipek, 2003). Besides family background, teachers' perceptions of value differences between themselves and parents also influence teachers' judgment of children's learning capacity (Hauser-Cram, Sirin, & Stipek, 2003).

Hauser-Cram, Sirin, and Stipek (2003) found that teachers have lower ratings of children when teachers perceive a difference in values held by parents, controlling for children's skills and family socioeconomic status (SES).

1.6.1. Children's Home Cultural Environment and Student-Teacher Relationships

Using data from a central-city urban southwestern school district, Farkas et al. (1990) conducted a study of cultural resources and social interaction in educational stratification. The study looked at differences in school achievements across gender, ethnicity, and SES groups by examining the informal academic standards that teachers used to reward more general skills, habits, and styles of students. The authors found that school rewards were based upon the teachers' judgment of student's non cognitive traits, such as study habits and appearance, as well as their cognitive performance. Students' cultural resources, represented by their skills, styles, and habits, served as signals; teachers, as gatekeepers, perceived such signals and conferred appropriate rewards. Students' conduct was in turn shaped by teachers' rewards. Other studies have conceptualized home cultural environment and skills that child can bring from home to include parents' having difficulty helping with homework (Smrekar, 1999), the sense of confidence and entitlement students feel when interacting with teachers (Lareau & Horvat, 1999), how comfortable students feel approaching teachers (Blackledge, 2001); language styles used at home, clothing styles, and styles of interaction between students and teachers (Carter, 2003).

These all studies have measured children's home cultural environment in different ways, but they all taps on the evaluative standards that teachers use to evaluate students beyond students' academic achievement. Research on the impact of family background on student-teacher relationships emphasize limited resources at home and the lack of skills children may bring to school. The above-cited studies indicate that teachers' perceptions should also be taken into consideration when examining factors that may influence student-teacher relationships in general, and teachers' evaluations and educational expectations in particular.

To make use of the rich information on homeroom teachers' opinions of the importance of children's family background influence and their evaluations of children, an exploratory factor analysis was conducted to identify different dimensions of the teachers' evaluations of children and the teachers' views on the importance of the children's home environment. The identification of factors was based on the results of oblique rotation, because it was expected that the factors were correlated (Brown, et al., 2003).

1.7 Teachers' Perceptions of Home Influence

This study explores whether teachers' perceptions of the importance of children's home environment may serve as one of the pass-ways through which the family background may influence a child's schooling.

1.7.1 Home cultural environment

This factor includes the teachers' answers to questions on whether they think the following factors are problematic for the child's future:

a. whether the parents share the school's values on education;

b. whether the parents are illiterate;

c. whether the parents are able to make good study plans for their children; and

d. whether the parents care about their children's schooling.

The construction of this factor reflects what teachers consider as valuable cultural resources from home that could facilitate children's schooling, and the importance of these resources.

1.7.2 Family social status

This reflects teachers' concern that a child's future may not depend only on their education, but also on the family's social network and social status. The factor includes the teachers' answers to questions about whether they think any of the following is important for the child's future:

a. the parent is in cadre position;

b. the family has a wide social network; and

c. the parents are able to locate good jobs for the child.

1.8. Teachers' Evaluations of Children

In this study, the teacher-student relationship is measured by the teachers' evaluations of children's academic competence and behavior at school, and their expectations of the children's future educational attainment.

1.8.1. Being a good student

The factor for a teacher's evaluation of a child being good student includes the teacher's rating of the child's learning capacity and achievement levels in language and mathematics comparing with other children. It also includes teachers' ratings of child's study habits, whether the child finishes homework, whether the child makes efforts to achieve better grades, and if the child participates actively in the classroom.

1.8.2. Experiencing Problems

The factor for the teachers' evaluation that a child is experiencing problems includes items that tap the teachers' assessment of the child's behavioral engagement at school. Questions include whether the child has disciplinary problems, is passive and/or disengaged in class, has problems interacting with other children, and seems to have already given up on school. The scale also includes the teachers' assessment of whether the child has any emotional adjustment problems, including whether the teacher thinks the child is nervous, lacks confidence, or is unhappy or depressed.

1.9 Child Family Background

Child family socioeconomic status is measured by the mother's education, father's education, as measured by years of formal schooling completed, and family wealth. In a rural setting, people do not have access to much cash income, and income from farming varies from year to year. Family wealth is a more stable measure of the economic situation than income. Family wealth is calculated by summing up the value of family property and assets, including housing, farm machinery and equipment, and household durable goods. The logged family wealth was used in multivariate analysis. The findings from this study can be summarized into two main points: first, net of objective measures of children's family SES and achievement at school, teachers' perceptions of the importance of children's home environment are closely associated with teachers' evaluations of children's academic competence and behavior in school; furthermore, teachers' perceptions of home importance are closely associated with teachers' educational expectations, both directly, and indirectly through teachers' evaluations of children. Second, teachers' expectations of children's future school attainment at an early point in time are significant predictors of children's later school persistence.

These results are in consistence with previous findings in research on student-teacher relationships: the cultural resources from children's home have an impact on their school experience through its influence on student-teacher relationships. What is innovative about this study is that it incorporates teachers' voices when examining the influence of the children's home environment on teacher-student relationships? The unique measurement of teachers' perceptions of the importance of children's home cultural environment and social status on children's schooling brings out the subtle influence of family background, which is often missed if only the conventional measures of families' socioeconomic situation are used. The interesting finding—that families' economic situation has almost no impact on teachers' evaluations and expectations, but that teachers' perceptions of the importance of home influence are closely associated with both teachers' evaluations and expectations—points to the importance of bringing teachers' attitudes and perceptions into research on teacher-student relationships at school.

The finding that teachers' expectations at early time point can help to predict children's school persistence later may indicate that when teachers have high expectations for children they are likely to provide more support and guidance to the children, which results in better chances for children to stay in school. It could also indicate that teachers predict accurately children's potential for further schooling. One thing is certain: teachers play an important role in shaping children's school experience, which are closely connected to children's school persistence, especially in the setting of rural China, where resources from children's home are very much limited. Children would benefit from closer connection and better understanding between their teachers and parents.

1.10 Research Objectives

1. To identify alienation among public school students.

2. To identify alienation among convent school students.

3. To examine the difference between mean scores of Public and Convent School students on alienation.

4. To examine the difference between mean scores of Public and Convent School students on School Environment.

5. To examine the difference between mean scores of Public and Convent School students on Home Environment.

6. To identify the influences of public school and convent school students on alienated adolescents.

1.11 Research Questions

1. Does alienation exist in Public school student?

2. Does alienation exist in Convent school student?

3. Does school environment influence Public School alienated students?

4. Does school environment influence Convent School alienated students?

5. Does home environment influence Public School alienated students?

6. Does home environment influence Convent School alienated students?

7. Is there any difference exist between Public and Convent school students on alienation?

8. Is there any difference exist between Public and Convent school students on School Environment?

9. Is there any difference exist between Public and Convent school students on Home Environment?

1.12. Operational Definitions

1.12.1 Alienation

Alienation is "the state or experience of being isolated from a group or an activity to which one should belong or in which one should be involved" (Mann, 2001).

1.12.2 School Environment

School environment is an external protective factor, which in the present research, is defined as an environment, where an adolescents student experiences caring relationships with and health expectations from the school faculty and takes meaningful participation in the school and class related matters.

1.12.3. Home Environment

Home environment is defined as an environment where an adolescent member experiences caring relationships with and healthy expectations from the family members and indulge in meaningful participation in family related matters.

Chapter 2
REVIEW OF LITERATURE

This chapter attempts to present a brief resume of research findings related to *Alienation of students in Public and Convent school students in relation to School Environment and Home Environment"*. The present review of literature examines the studies which lead to the formulation of the problem of present study.

2.1. Alienation

Adolescent alienation is a problem for some young people and may lead to behaviors such as vandalism, violence, and gang activity (Brown, Higgins, Paulsen, 2003; & Hawkins et al., 2000). Many variables can trigger an adolescent's feelings of isolation and disconnection such as changes in family circumstances, lack of social acceptance, and academic underachievement (Hawkins et al., 2000).

Seeman (1959) undertook one of the first studies in social psychology that aimed to clarify the traditionally sociological concept of alienation by examining its multidimensionality. He identified five dimensions: self-estrangement, powerlessness, social isolation, meaninglessness, and normlessness.

The debates surrounding the concept of alienation are numerous and raise important questions. For example, is it a state, a trait, or a self-regulating process? Is it unidimensional or multidimensional? (Dean, 1961; Mackey & Ahlgren, 1977; Mau 1992; Roberts, 1987; Williamson & Cullingford, 1998).

Newman (1981) observed that perception does not, however, excuse school improvement and does not justify abandoning the effort to create less alienating schools. So long as there is a possibility to improve school life for all students, schools, and educators have a moral obligation to do so . In order to create less alienating schools, educators, administrators, and parents must identify the characteristics (e.g., unfair practices, policies, procedures, treatment of students by teachers and other school personnel) within schools and within the students most at risk (e.g., students with disabilities,) that exacerbate the feelings of alienation experienced by these students.

Roberts (1987) addresses the methodological issues related to the conduct of alienation. In his investigation of the multidimensionality and structure of alienation, data from a multicultural sample of 3,101 male workers from the United States, Poland, and Japan were analyzed using second-order confirmatory factor analysis. The results clearly improve our knowledge regarding the structure of adult alienation. While confirming multidimensionality, Roberts (1987) also argues that self-estrangement and powerlessness are most closely related to alienation. Further, his statistical model suggests that manifestations of alienation are similar to the original Marxian conceptualization.

The present study investigated the theoretical structure of adolescent alienation by applying the same structural equation modeling (second-order confirmatory factor analysis) used by Roberts (1987). The manifestations of alienation were examined in a sample of boys and girls to determine whether the constructs measured are invariant across gender. This is considered an important step before engaging in further exploration of the risk factors and consequences of alienation.

Margalit (1991) concluded that four clusters of students existed, including: lonely, aggressive students who were disruptive and viewed themselves as lonely, (2) lonely, nonaggressive students who viewed themselves as very lonely and were rated by their teachers as having low levels of disruptive behaviors, (3) nonaggressive and non-lonely students did not view themselves as lonely, nor did their teachers rate them as disruptive, and (4) extremely lonely students who emphasized their lonely feelings, but were not rated as disruptive. Results from the Margalit (1991) study indicated that loneliness was significantly predicted by a student's peer acceptance, computer activities, and social skills. All three were negatively related in that the lower the ratings on peer acceptance, computer activities, and social skills, the greater the loneliness scores of a student.

Mau (1992) developed a 24-item scale to measure Seeman's dimensions of alienation. Based on the preliminary work of Mackey and Ahlgren (1977) with students, and using multidimensional scaling with a sample of 2,056 adolescents from three Hawaiian high schools, Mau (1992) validated an alienation inventory that comprised four subscales: powerlessness, meaninglessness, normlessness, and social estrangement (i.e., social isolation).

Williamson and Cullingford (1998) corroborated the multidimensionality of Mau's scale with a sample of 254 adolescents from England, using orthogonal exploratory factor analysis. While both studies confirmed the multidimensionality of this alienation scale, they used orthogonal analysis, which unfortunately does not allow for an elucidation of the relationships between the dimensions. This is an important limitation, since clarifying these relationships is crucial for understanding the structure of the global construct, namely alienation.

Seeman's (1959) original five dimensions were slightly modified by Mackey and Ahlgren (1977) and Mau (1992), applied to adolescent populations in school contexts. They can be described as follows. Self-estrangement has its roots in classical philosophy and taps humanistic ideas mostly related to a discrepancy between actual and idealized self. This dimension manifests itself in adolescents who have low self-esteem and feel bored with life, in which they perceive no purpose. Powerlessness reflects fatalism, pessimism, and a perception of losing control over one's own life. This dimension is similar to the psychological notion of external locus of control. Social isolation is salient to youths who perceive a lack of intimate relationships, such as with friends, thus leading to a feeling of loneliness. Normlessness can be defined as a belief that socially disapproved behaviors may be used to achieve culturally defined goals. For example, adolescents who strive for good grades in school but perceive that they are not provided with the means to achieve them may cheat during exams. Meaninglessness, which is an important dimension to explore within an educational context (Williamson & Cullingford, 1998), characterizes youths who perceive little or no relationship between what they learn in school and what they will do in the future. The two latter dimensions clearly integrate Durkheim's and Merton's theories regarding anomie into Marx's alienation theory.

Seeman's dimensions are derived from very different theoretical frameworks and, as mentioned previously, the logical relations between them are not clear. Conceptually, if these dimensions are related to a single general construct, namely alienation, they would be expected to share a considerable amount of variance. This covariance would suggest that these dimensions possess common causes and consequences. On the other hand, if these dimensions are too closely related, it may be more useful to consider them individually and to build theory around each one. Mackey and Ahlgren (1977), Mau (1992), and Williamson and Cullingford (1998), have empirically explored the multidimensionality of the alienation construct in youth.

Sometime student's early years in school develops an attitude about school and education in general. For some, this attitude may be positive and inclusionary in nature, while for others this attitude may be one of estrangement or isolation. If the latter attitude is adopted by the student, the result could be increased gang membership (Calabrese & Noboa, 1995; Shoho, 1996), destructive behaviors (Ascher, 1982; Staples, 2000), and/or dropping out of school.

Arnett (1996) viewed adolescents' preference for heavy metal music as being related to alienation. Unfortunately, few of these studies offered an adequate empirical definition of alienation before integrating the concept into their statistical models.

Mackey and Ahlgren (1977), Israel (1972), and Williamson and Cullingford (1997) have observed that throughout the 20th century, the concept of alienation has received substantial attention in the social sciences. Since its introduction by Hegel and Marx, this concept has been defined in a variety of ways that reflect the various disciplines and specific views of researchers who study it. Alienation research peaked during the 1970s and has received declining attention until recently. Many have criticized this concept as being too broadly used to describe nearly any kind of aberrant behavior, ranging from political manifestations to psychopathology Nonetheless, there has been a recent upsurge of interest in applying the concept to adolescents' experiences, most specifically their scholastic experience.

Theoretical and measurement issues still need to be clarified. Since 1960s a number of researchers have tried to gain a better understanding of the empirical definition of alienation and its manifestations in adolescent boys and girls (Arnett, 1996; Dean, 1961; Mackey & Ahlgren, 1977; Mau, 1992; Roberts, 1987; Williamson & Cullingford, 1997, 1998).

During the past 40 years, empirical studies have tried to determine the psychological manifestations of alienation, mostly in adults but also in adolescents. In psychological and educational studies, for instance, adolescent alienation has been correlated with externalized behaviors such as drug use (Jessor & Jessor, 1977), truancy, delinquency (Williamson & Cullingford, 1998; Calabrese, 1990), and suicide (Wenz, 1979; Young, 1985). It also has been associated with internalized problems such as low self-esteem (Williamson & Cullingford, 1998), psychological distress, and depression (Abdallah, 1997).

Tomaka and Palacios (2006) demonstrate fairly consistently that social alienation is related to negative health outcomes. Their study further indicates a significant relationship between a sense of alienation and health status.

Rayce et al. (2008) conducted a study in Denmark focused on "aspects of alienation and symptom load among adolescents." The study demonstrates an association between alienation and experiencing daily physical and psychological symptoms. In their study, the symptom load was split into two dimensions, physical (headache, stomachache, back pain, and dizziness) and psychological (feeling low, irritated/bad temper, nervous, and sleeping difficulties). They suggest that more research is needed to identify the pathway between alienation and symptom load.

Crinson and Yuill (2008) in their theoretical study about alienation and inequalities in health investigated how alienation theories can be applied with respect to health outcomes. Based on previous research they attempt to determine how a sense of alienation can be associated with poor health and health outcomes such as stress, emotional deadness, psychological and physical well-being, aggression, and dissatisfaction.

Alienation in adolescence may lead to behaviors such as truancy, decreased attendance, academic failure, vandalism, violence, and gang activity. Many variables such as changes in family circumstances, lack of social acceptance, academic underachievement, and oppressive or inequitable practices can trigger an adolescent's feelings of isolation and disconnection from family, friends, and the community. (Hawkins, et al., 2000; McInerney, 2009).

Discussing the nature of schizophrenia, De Block and Adriaens (2011) noted, "the critical error of dualistic understandings of mind and its pathologies is this: individual experience is stripped of its interpersonal, social, and existential dimensions". A similar statement could be made in a psychological construct with little consideration for the possible spiritual factor.

The problem of spirituality was rarely identified as a factor in recent social alienation research, the origin of which appears rooted in the Western dualist concept of the human being. Both Plato and Aristotle argued that the mind was immaterial, a belief embedded in the Western mind by Descartes and continued by Freud, Husserl, and Jaspers (as reported by De Block & Adriaens, 2011).

2.2. School Environment

Adolescents spend a significant portion of their time in school, and "it is the role of the school to see that they are prepared for society and future employment" Yet many students such as students from diverse ethnic and linguistic backgrounds and students with non-traditional family circumstances are disengaging from school or completely dropping out without being appropriately prepared socially or academically for future responsibilities (Brown et al., 2003; Calabrese, 1987, 1990; Gullota, 2001; Mau, 1992; Oerlemans & Jenkins, 1998).

Sizer (1984) reported that merely reducing school size, however, is not sufficient on its own to achieve positive academic and social outcomes for students. Structurally, small schools often foster close student- teacher relationships. Evidence in child development and school counseling research support the belief that students learn best in a close-knit, nurturing environment in which no child can "fall through the cracks" of the large, impersonal, bureaucratized high school.

Teachers and administrators charged with students' academic and social growth are themselves deprived of the conditions and resources that support their own capacity to support students (Sizer, 1984). The isolation of the teachers from each other—even when working with the same students—denies them opportunities to discuss individual students' needs, develop appropriate curriculum, or solve problems. Teachers are asked to work with large numbers of students, cover an ever-expanding curriculum, oversee extracurricular clubs and activities, and raise their students' test scores.

Geary (1998) interviewed 35 African-American sophomores and juniors at an all Black inner city high school. Findings suggested that school success was facilitated by a teacher who was perceived as caring, available, understanding, encouraging, respectful, listening, and having a sense of humor.

In a larger study, Coburn and Nelson (1989) surveyed more than 300 Native-American high school students in Washington, Montana, Oregon, and Idaho. These students cited influential teachers as having the following characteristics: respectful, caring, listening, conveying a positive attitude, providing assistance readily, encouraging, available, engaging the student in the learning process, and affirming students positively.

If school personnel are going to create safe, healthy, and inclusive environments for students with disabilities, they must rid the school environment of all factors that may contribute to their alienation (Fetco, 1985; Wilson, 1989).

The typical high school, however, is organized in such a way that creates a false dichotomy between social and academic development. The fragmented structure of high schools (Powell, Farrar, & Cohen, 1985) rarely makes place for sustained consideration of the social, psychological, and emotional issues affecting young people, locating such issues strictly within the purview of guidance counselors.

In a study of Catholic high schools for disadvantaged urban youth, Bryk et al. (1993) maintain that strong "academic press" coupled with intense social relationships were key components of the schools' effectiveness with their students.

Bryk et al. (1993) believed that one common component of a distributed counseling program is "advisory," in which each teacher (or other staff member) serves as an advisor to a group of students, often smaller than typical class size, that meets several times a week to provide the social supports.

McMillan and Reed (1994) agreed that "positive experiences in school help provide students a sense of belonging, bonding, and encouragement". And "often make the difference between positive school experiences and frustration or alienation" (Chaskin & Rauner, 1995).

Furthermore, Coley (1995) analyzed data from the National Center for Education Statistics to identify factors that contributed to the drop-out rate in the United States. He reported that more than 43% of students left school because they disliked school; less than 40% cited getting unsatisfactory grades; more than 25% stated strained relationships with teachers; and less than 25% lacked a sense of belonging in school.

Cotton (1996) observed that Students in small schools are more likely to be involved in extracurricular activities and to hold important positions in school groups than are similar students at large high schools .

Cotton (1999) found that advantages of a small high school environment are not merely academic. There is significant evidence that small learning environments foster closer student-teacher relationship, leading to many social, emotional, and psychological benefits for young people.

The phenomenon of alienation describes a student's level of academic and social disengagement from school. Behaviors associated with student alienation are numerous and include hostility, passivity, withdrawal, poor quality work, disinterest, lack of involvement and initiative, suspensions, expulsions, and non-completion In its extreme, feeling alienated from the school environment can contribute to violence and suicidal ideation (Hawkins et al. 2000; McMillan, 1992).

A caring demeanor is critical, "especially for culturally diverse students who may be at risk of failing or who may be disengaged from schooling" (Perez, 2000). Wasley et al. (2000) concluded that small schools also report proportionally fewer disciplinary problems and incidences of violence compared to large schools.

Darling-Hammond (2002) found that students at smaller high schools manifest more positive attitudes about being in school, as well as less of a sense of alienation than students in larger schools

Moreover, teachers are not provided with enough resources and professional development to realize such objectives (Darling-Hammond, 1997, 2002). Given these realities, it is little wonder that teachers are typically unprepared to deal with issues of social and emotional development and limit to their professional activities to the academic sphere.

Brown et al. (2003) describes that alienation demonstrates estrangement and disaffection from the educational institution and the worth of institutionalized education. A student's level of identification with and participation in the goals of schools has been impacted by a loss of or inability to form bonds with peers and school adults.

Recent high school reform efforts have focused on the creation of small schools (Ancess, 2003). Proponents point to the academic benefits that students gain from small learning communities such as improved academic performance, higher graduation rates, and lower dropout rates (Center for Collaborative Education, 2003). In general, students in small schools report a greater sense of belonging, leading to more positive social behaviors (Center for Collaborative Education, 2003).

Greene and Winters (2005) found gender disparities in discipline and non-completion statistics have focused attention on boys. While these issues for girls are certainly as important, the literature suggests that boys' education is being impacted to a degree that warrants concern. Research indicates that males have a higher probability of becoming non-completers than do females (Voisin et al., 2005). Caring relationships were especially viewed as a positive intervention approach for students who may turn to at-risk behaviors

Johnso (2005) viewed that just what constitutes alienation in the school context is problematic, but it undoubtedly involves varying degrees of student estrangement from the learning process, as manifested in behaviors such as passive resistance, withdrawal of labor, truancy, disruptive activities, violence, self harm and dropping out of school.

Knesting (2008) interviewed 17 students (African-American and White) in grades 9-12 (ages 15-19) who were at risk of leaving school. Findings revealed that students were more likely to stay in school when they perceived teachers as caring. Teachers caring were demonstrated by behaviors that enhanced students' potential, fostered their self-esteem, valued their opinions, and respected them as individuals. Other studies have suggested interventions to help gifted students stay in school such as improving student-teacher relationships, by building students' self-esteem, conveying a positive attitude, and validating them as individuals

These students, who too often come from culturally and linguistically diverse backgrounds and non-traditional family circumstances, disengage from school and, consequently, may be ill-prepared for future responsibilities (Brown et al., 2003; Education Trust, 2002; Gullotta, 2001, Johnson, 2009; Mau, 1992; Oerlemans & Jenkins, 1998).

According to Constantine, Kindaichi, and Miville (2007) preparing the social environment to ensure a more welcoming atmosphere by implementing psycho educational opportunities such as classroom guidance could support student-school connections. Additionally, creating a "welcome wagon" comprised of culturally and linguistically diverse students to accompany new students for the first several days and introduce them to their teachers and the unspoken rules of the cafeteria, hallways, playground, and parking lot could quickly reduce anxiety and support student belonging. Opportunities such as newcomer lunches, after school meet-n-greets, and short weekend excursions could involve students, teachers, and parents into the process of creating a welcoming, inclusive culture.

Hansen and Toso (2007) observed some experts acknowledge that risk factors such as poverty, teen pregnancy, substance and physical abuse, peer pressure, and family support may contribute to student disengagement at school, institutional practices such as segregation by ability, an ineffectiveness of teachers to connect with students, and students' perceptions that teachers are uncaring individuals also contribute to students' negative disposition towards learning. In fact, one of the reasons students leave school before graduating, is boredom and an unchallenging curriculum.

Thompson (2007) invited teachers and high school students to respond to questions about caring. While most teachers reported caring about their students, nearly 40% of the student respondents were in disagreement, illustrating the complex nature of caring. "As many students continue to face negative experiences and a lack of caring and meaningful relationships with school personnel, their motivation to attend school becomes futile and too often just give up (Epstein, 1992; Hansen & Toso, 2007).

In contrast, Hansen and Toso (2007) surveyed 14 gifted students who had dropped out of school. These students reported the following reasons for not graduating: no sense of belonging, limited positive bonds with educators, undemanding class instruction, and feeling disrespected and misunderstood.

According to McInerney (2009), Bridgeland et al. (2006) Moore and Friant (2010) considerable evidence exists that a student's experience of alienation is profoundly shaped by psychological and personal factors yet school factors, including inequitable scheduling, grading, and discipline policies, may also contribute to a student's feeling of disconnection. School factors that have been linked with increased alienation include students' perception of lack of control in the school environment, degree of social acceptance, teachers' low expectations, and relevancy of the school curricula, toxic school culture, and discriminatory practices.

Students new to a school system will connect as they perceive likeness with other students (Booker, 2007). While not all school success correlates to belonging or connectedness, schools that offer programs grounded in the building of relationships can counteract disengagement leading to alienation. Price (2008) found that while it is common to read about the success of reform initiatives implemented in schools across the nation, the dropout rate has increased at an alarming rate.

According to the National Institute of Educational Studies (NCES, 2010), one of the primary reasons students leave school before graduation is a sense that school is not for them. Several other studies support the potential significance of alienation in poor academic performance and non-completion. The statistics on boys' increasing disengagement from academics provided by the National Institute of Educational studies NCES explain report and their educational experiences are evidence of this concern. These recent statistics show boys are more likely to repeat a grade, to require special education services, or to be diagnosed with either attention-deficit disorder or hyperactivity. Eighth-grade boys are 50% more likely to be held back a grade than girls. By high school, 67% of all special-education students are boys. Boys receive 71% of all school suspensions. The report identifies four factors that highlight the possible reasons boys may experience schools differently than gifts. They are: (a) the influence of socioeconomic background, (b) different attitudes and behaviors with respect to school and learning, (c) the effects of stereotypes, and (d) the influence of peer groups. The opportunity to create an operational policy that can support boys' connection to the core mission of schools is enhanced by gaining an understanding of their lived experiences. Their perceptions lend credibility to reports such as the NCES, and are achieved by giving voice to youth that identify as disengaged both academically and socially.

It was believed that the dynamic process of alienation requires more than one negative incident or one "bad" teacher. Students reach the point of non-completion after years of feeling less familiar, less considered, less teachable, and less likeable (Edwards & Mullis, 2001; Johnson, 2009). Alienation is attributed to both school and home environments; participants described specifically how each impacted their ability to achieve academically and socially. Schools that give priority to creating welcoming and accepting climates can positively impact students' sense of belonging and academic self-efficacy.

2.3. Home Environment

The family is the primary social system for children. Rollins and Thomas (1979) found that high parental control were associated with high achievement. Student voices need to be heard in order that deeper issues of power can be considered because each individual has privileged knowledge of their own selves (Delpit, 1988).

Rogoff (1990) used three lenses to understand the development of individuals within a socio-cultural framework. These lenses allowed one to focus on an individual, their relationship with others and the cultural environment in which they are immersed. In this study, these lenses have been useful for helping to uncover those areas of privileged knowledge held by students and the tensions between institutional practices and the students' best interests. Students who have experienced transition between home education and formal education have also been silenced by the failure of mainstream educational researchers to see home education as a worthwhile research topic. Student views and experiences are a valuable way to enrich mainstream understanding, and open otherwise closed topics for discussion.

Cassidy and Lynn (1991) included a specific factor of the family's socioeconomic status, crowding, as an indicator of how being disadvantaged affects educational attainment. They found that a less physically crowded environment, along with motivation and parental support, were associated with higher educational levels of children. Religiosity as an aspect of the family environment is another independent variable possibly influencing academic achievement (Bahr, Hawks, & Wang, 1993).

Cassidy and Lynn (1991) explored how family environment impacts motivation and achievement. This means that motivation served as a mediating variable between home background, personal characteristics, and educational attainment. Specifically, female students' academic performance is less associated with their interests than male students' academic performance (Schiefele, Krapp, & Winteler, 1992).

Ferguson (1991) and Peng and Wright's (1994) analysis of academic achievement, home environment (including family income) and educational activities, concluded that home environment and educational activities explained the greatest amount of variance. In conclusion denying the role of the impact of a student's home circumstances will not help to endow teachers and schools with the capacity to reduce achievement gaps (Hammer, 2003). Niebuhr's (1995) findings suggest that the elements of both school climate and family environment have a stronger direct on academic achievement.

Wang, Wildman and Calhoun (1996) argued that parental influence has been identified as an important factor affecting student achievement. Results indicate that parent education and encouragement are strongly related to improved student achievement.

Phillips (1998) also found that parental education and socioeconomic status have an impact on student achievement. Students with parents who were both college-educated tended to achieve at the highest levels. Income and family size were modestly related to achievement. Social development was reviewed in a number of Australian studies (Barratt-Peacock, 1997; Brosan, 1991; Clery, 1998; Krivanek, 1988; Simich, 1998; & Thomas, 1998). The Australian findings in this area also support the findings given in overseas literature which indicate that home educated students are generally socially well adjusted and have a healthy self-esteem. Howse (1999) found that academic achievement is accomplished by the actual execution of class work in the school setting. It is typically assessed by the use of teacher ratings, tests, and exams.

Research shows that student's perceptions of academic competency decline as they advance in school (Eccles, Wigfield, & Schiefele, 1998). Schunk and Pajares (2002) attribute this decline to various factors, including greater competition, less teacher attention to individual student progress, and stresses associated with school transitions. Students were motivated by teachers who cared about student learning and showed enthusiasm. These teachers introduced topics in an interesting and challenging way, used varied teaching strategies, and promoted student involvement by allowing participation in the selection of learning activities (Cothran & Ennis, 2000).

According to Hammer (2003), the home environment is as important as what goes on in the school. Important factors include parental involvement in their children's education, how much parents read to young children, how much TV children are allowed to watch and how often students change schools. Achievement is not only about what goes on once students get into the classroom. It's also about what happens to them before and after school. Parents and teachers have a crucial role to play to make sure that every child becomes a high achiever.

There have been few Australian studies of home schooled student's academic achievements. The little research work indicated that student academic achievement was generally average to well above average (Harding, 2003; McColl, 2005; Simich, 1998; Thomas, 1998).

Student views of their home education experience have also been researched in several Australian studies (Clery, 1998; Honeybone, 2002;

McColl, 2005). Students have generally reported positively about their home education experience. The few comments made about missing peers at school were generally tempered by the benefits of being able to learn flexibly at their own pace. Self-esteem appeared to be healthy among these students with a number of them crediting their own healthy self image directly to their home education experiences.

Most overseas literature on home education and transition experiences with formal institutions has been generated in the United States of America with some work having been conducted in Great Britain. Five of these studies focused on home educated student entrance to tertiary level institutions or the work They found that students' adjusted well academically and socially to post secondary institutions or the workforce and often better than many of their formally educated peers (Geray, 1998; Goymer, 2001; Lattibeaudiere, 2000).

There have also been overseas studies focusing on interactions between home education and formal primary or secondary schools (Four of these studies discussed various aspects of the interplay between home schooling and formal schooling, especially from an administrative point of view (Adams, 1992; Davis, 2000; Fairchild, 2002; Hanna, 1996; Krout, 2001; Stoppler, 1998). Topics ranged around attitudes of superintendents, home schooler access to school resources and partnerships between public schools and home educators indicating that home schooled students were moving between the two systems and sometimes making use of both at the same time.

There were three overseas studies (Krout, 2001; Snyder, 2005; Stoppler, 1998) which examined the movement of students between home schooling and formal education. Stoppler (1998) indicated that parents returned children to formal schools because they believed that it was in the best interests of the child even though these needs varied between families and children. Snyder (2005) found that home educated students generally adjusted well into formal schools both academically and socially. In a study of the entrance of children from three families into formal education, it was found that these families sent their children to school to encourage the development of social skills with professionals and peers.

In Australia, there have been several studies on why parents choose home education (Barratt-Peacock, 1997; Habibullah, 2004; Harding 1997;

Harp, 1998) how they conduct their home schooling programs (Jacob et al. 1991; Barratt-Peacock, 1997; Habibullah, 2004; Simich 1998; Thomas 1998) and the law (Education Queensland 2003; Harding & Farrell 2003; Jeffrey & Giskes 2004). However, there do not appear to be any studies which have looked at the movement of students into or out of formal education. This study explores the experiences of three such students, all of whom have moved into or out of formal education and engaged in home education at some point of their student life.

Socio-cultural theory as described by Rogoff (1990, 2003) has been used to inform this study. She discussed the need to assess situations from various perspectives. The student point of view is perhaps the most important aspect of the transition experience. Other aspects of these students' stories are found by looking at their environment. These include students' views of their families and the institutions with which they have interfaced. Institutional misunderstanding of student motives has, in the past, led to conflict creating practices by institutions which have further alienated struggling students (Hedegaard, 2005).

Hughes, Gleason, and Zhang (2005) found that teachers' perceptions of student teacher relationships and parent-teacher relationships also add variations to teachers' evaluations of children's learning capacities in addition to children's measured achievements. Based on different social structures, it is obvious that measures for studying home environment reflect the society they are intended to measure in their geographical area.

Regarding the home literacy environment, measures have included examining the extent of availability of educational materials in the home, various teaching strategies provided by parents and other people in the home and neighborhood (Sénéchal & LeFevre, 2002).

In this chapter, it is aimed to provide an overview of the literature on school environment and home environment. The literature review indicates that emotional empathy is linked to various variables which include both personal characteristics and personality variables. The investigators have not come across a single study where the relationship between spiritual personality and emotional empathy was examined. **Chapter 3** presents the methodology of the present study.

Chapter 3
METHODOLOGY

The present investigation was carried out to examine *"Alienation of students in Public and Convent school students in relation to School Environment and Home Environment"*. The concept of methodology includes four aspects, namely, subject, tools, procedure and data analysis. These four aspects of overall research methodology can be taught of as forming a case for execution of present study. Additionally, the methodology provides detailed information about how the subject used for the study, the description of the participants and the measured used in the study.

Research may be described as a systematic pursuit of new knowledge. Its aim is to extend the frontiers of knowledge, to bring about an increase in the existing fund of knowledge. Research comprises of defining and redefining problems, formulating hypothesis or suggesting solution, collecting information, organizing and evaluating data, making deductions and reaching conclusions and at last testing carefully.

Formulating of research questions along with sampling weather probable or nonprobable is followed by measurement that includes surveys and scaling. This is followed by research design, which may be either experimental or quasi-experimental.

3.1. Research Design

The main function of research design is to provide information for the collection of relevant evidence with minimal expenditure of effort and time. It depends mainly on the research objectives and questions.

3.2. Subjects

The data sample for the present study consists of 300 adult students. Subjects were drawn from Al-Barkaat Public School and convent Our Lady of Fatima School of Aligarh (UP), India. Subjects were classified on the basis of scores obtained on alienation scale on which Quartile 1 and Quartile 3 were calculated to get the alienated subjects. Public school subjects whose scores fell below Q1 (45) were identified as low alienated and subjects whose scores fell above Q3 (52) were categorized as high alienated. Convent school subjects who have obtained scores below 41 were categorized as low alienated and those subjects whose score were above 49 they categorized as high alienated subjects. The sample further classified into two categories i.e. 50 students were drawn from convent schools and other 50 were taken from public schools.

3.3. Tools

The following tools were used in the present study:

3.3.1. Alienation Scale

Alienation scale was developed by Kureshi and Dutt (1979).This scale comprised 21 items with four alternative response categories (always, often, sometimes, never, representing the five factors despair, disillusionment, unstructural universe, psychological vacuum and narcissum. In order to get reliability of alienation scale Cronbach's Alpha was calculated on 150 students with the help of Statistical Package of Social Sciences (SPSS). It was observed that the Cronbach's Alpha of alienation is 0.76 which is quite high; which clearly indicates that this scale is reliable for the present study.

3.3.2. Home Environment and School Environment Scales

The Home and School Environment questionnaires are subscales of The Resilience and Youth Development Module (RYDM) which is a component of California Healthy Kids Survey (WestEd, 2002). The RYDM is devoted completely to assessing the internal and external assets associated with positive youth development and resilience (WestEd, 2002). The RYDM provides comprehensive and balanced coverage of external assets in home and school environment.

The full RYDM contains 59 questions that measure 17 external and 6 internal assets in the home, school, community, peer group and in the individual domains. Both, the home and school environment scales for measuring external assets, have 9 items each, and measure three common dimensions:

1. Caring relationships i.e. supportive connections with others, like having a person who is there and who listens non-judgmentally.

2. High expectations i.e. the consistent communication of message that the adolescent student/family member can and will succeed, a belief in youth's innate resilience, and the provision of guidance that is youth centered and strengths focused.

3. Meaningful participation i.e. the involvement of adolescent student/ family member in relevant engaging and interesting activities and having the opportunities for responsibilities and contributions.

Each item has four response options (very much true, pretty much true, a little true and not at all true) out of which the participants had to choose the one option which they felt best applied to them. The scoring ranged from 4 to 1 on the four point Likert scale. The value 4,3,2,1, attached to each response option were averaged and then the following score categories were derived.

High scores –are percentage of students with average item response is above 3. High score of the two scales were indicators on good home and school environment.

Moderate score – are percentage of students with averaged item response of at least 2 but less than or equal to 3. Moderate score on the two scales were indicators of moderate home and school environment.

Low scores— are percentage of students with averaged item response below 2. Low scores on the two scales were indicators of poor home and school environment.

Cronbach's alpha of the three dimensions in the home environment scale ranged from .70 to .80. For the three dimensions of school environment scale, Cronbach's Alpha was found to range between .75 to .90 (Constantine & Benard, 2001). Good construct validity of both the scales has also been reported (Hanson & Kim, 2007).

3.4. Procedure

First of all, the investigator obtained permission from the Principles of the both the schools Al-Barkaat Public School & Our Lady of Fatima School Aligarh (UP). The investigator established rapport with the respondent students, prior to administer the proposed scales, and assured them that their responses would be kept strictly confidential and would be utilized for the research purpose only. After establishing rapport with the students the data were collected in their classrooms in many sessions.

3.5. Statistical Analysis

Data were analyzed by using Statistical Package of Social Sciences (SPSS) version 12.0. In order to answer research questions set in chapter one, the present investigator used Stepwise Multiple Regression Analysis (SMRA) and t-test also applied to examine mean differences between public and convent school's students.

Chapter 4
RESULTS AND DISCUSSION

The entire data of the study after tabulation were analyzed by using Stepwise Multiple Regression Analysis (SMRA) and t-test.

Firstly, total sample was were taken to study the significant influence of school environment, home environment and their dimensions on students' alienation, and then public school students and Convent school students were taken separately. Thereafter, public and Convent school students and male and female students were compared on alienation, school environment, home environment and their dimensions.

Table–4.1 highlights the model summary of SMRA showing the significant predictors of alienation of high school students for the total sample irrespective of the type of schools. Table clearly shows that only one dimension of home environment namely 'Expectation in home' out of eight predictor variables emerged as significant predictor of alienation which accounted for 4.6% (R^2= .046) of variance in alienation. It is also evident from the Table–4.1A that 'Expectation in home'- a dimension of home environment having significant influence on alienation among students as its F value is F=4.776, which is found significant beyond .05 level of confidence.

Table - 4.1: Model Summary of Stepwise Multiple Regression Analysis Showing the Significant Predictor Variables of Alienation on Total Sample N= 100

Model	R	R Square	Adjusted R Square	Std. Error of the Estimate	Change Statistics		
					R Change Square	F Change	Sig. F Change
1	.216(a)	.046	.037	10.206	.046	4.776	.031

a. Predictors: (Constant), HE2-Expectation in home

4.1A: ANOVA

Model		Sum of Squares	df	Mean Square	F	Sig.
1	Regression	497.453	1	497.453	4.776	.031(a)
	Residual	10208.257	98	104.166		
	Total	10705.710	99			

a. Predictors: (Constant), HE2-Expectation in home

b. Dependent Variable: Alienation

4.1B: Coefficients

Model		Unstandardized Coefficients		Standardized Coefficients	t	Sig.
		B	Std. Error	Beta		
1	(Constant)	39.165	3.847		10.181	.000
	HE2-Expectaion in home	1.135	.519	.216	2.185	.031

a. Dependent Variable: Alienation

The analysis in its further step calculated coefficients of regression which were given in Table–4.1B, where under column standardized coefficients the beta (β) values of significant predictor was given. All the β values are measured in standard deviation units and the outcome is changed as a result of one standard deviation change in predictor variable. Standardized beta coefficient values indicate the relative influence of each significant predictor variables in influencing dependent variable. Table–4.1B shows that the β values for 'Expectation in home' is β= .261 which is found significant at .05 level of confidence. Beta value clearly indicates that 'expectation in home' significantly and positively related to alienation among students. It means that those students perceived high expectation from home (parents or sibling) were higher in their level of alienation as compared to students having low expectation from home.

Having explained the findings for the total sample irrespective of type of schools, the results highlighted from Table–4.2 to Table–4.2B are pertaining to the sub-sample groups of public school students. Table–4.2 shows model summary indicates significant predictive influence of predictor variables namely, 'Caring relationship in school'- a dimension of school environment and 'Home environment as a whole' on alienation of the students of public school, and 20% (R^2= .200) variance in alienation are explained by these variables, as their F values ranging from F= 5.888 to F= 6.644 were found statistically significant beyond .05 level of significance (Table–4.2A).

Table 4.2:Model Summary of Stepwise Multiple Regression Analysis Showing the   Significant predictor Variables of Alienation on Students of Public school (n= 50)

Model	R	R Square	Adjusted R Square	Std. Error of the Estimate	Change Statistics		
					R Square Change	F Change	Sig. F Change
1	.349(a)	.122	.103	8.859	.122	6.644	.013
2	.448(b)	.200	.166	8.542	.079	4.630	.037

a. Predictors: (Constant), SE1-Caring relationship in school
b. Predictors: (Constant), SE1-Caring relationship in school, Home Environment
c. Types of School = Public School

4.2A: ANOVA

Model		Sum of Squares	df	Mean Square	F	Sig.
1	Regression	521.397	1	521.397	6.644	.013(a)
	Residual	3767.103	48	78.481		
	Total	4288.500	49			
2	Regression	859.192	2	429.596	5.888	.005(b)
	Residual	3429.308	47	72.964		
	Total	4288.500	49			

a. Predictors: (Constant), SE1-Caring relationship in school
b. Predictors: (Constant), SE1-Caring relationship in school, Home Environment
c. Dependent Variable: Alienation
d. Types of School = Public School

4.2B: Coefficients

Model		Unstandardized Coefficients		Standardized Coefficients	t	Sig.
		B	Std. Error	Beta		
1	(Constant)	58.415	5.090		11.477	.000
	SE1-Caring relationship in school	-1.660	.644	-.349	-2.578	.013
2	(Constant)	44.498	8.119		5.480	.000
	SE1-Caring relationship in school	-1.811	.625	-.380	-2.898	.006
	Home Environment	.754	.350	.282	2.152	.037

a Dependent Variable: Alienation
b Types of School = Public School

Moreover, Table–4.2B of coefficients in its Step-2 under the column of standardized coefficients shows the beta values for 'Caring relationship in school'" a dimension of school environment was β= -.380 which is found significant at .01 level of confidence whereas, for 'Home environment as a whole' â= .282 which was significant at.01 level of significance. Beta value clearly shows that 'Caring relationship in school' negatively related to alienation whereas 'Total home environment positively related to alienation in public school students.

So far as 'caring relationship in school' is concerned, it refers to teachers and staffs caring attitude towards students and 'caring relationship in school' has significantly negative influence on alienation. The finding seems to be very logical as high caring attitude towards student's lead low alienation in students. .

Another significant predictor variable which is significantly related to alienation is that of 'Home environment as a whole' and the relationship between these two variable was found positive (Table–4.2B). This finding seems to be quite interesting and challenging in defending pattern of results as the high level of home environment lead high alienation in students.

After undertaking the sample of Public school students, Convent school students were taken for analysis and findings of the analysis have been shown in the tables from Table–4.3 to Table–4.3B. It is evident from the Table–4.3 of model summary that 'Meaningful participation in home'- a dimension of home environment emerged as significant predictor of alienation and it accounts for 14.5% of variance (R^2= .145) in alienation. The same is very clear from the Table–4.3A of ANOVA as its correspondent F value i.e., F=8.123, was found significant beyond .01 level of confidence. The result reveals that 'meaningful participation in home' found important so far as level of alienation among school children are concerned.

Table 4.3:Model Summary of Stepwise Multiple Regression Analysis Showing the Significant Predictor Variables of Alienation on Students of Convent School (n= 50)

Model	R	R Square	Adjusted R Square	Std. Error of the Estimate	Change Statistics		
					R Square Change	F Change	Sig. F Change
1	.380(a)	.145	.127	10.486	.145	8.123	.006

    a Predictors: (Constant), HE3-Meaningful participation in home
    b Types of School = Convent School

4.3 A: ANOVA

Model		Sum of Squares	df	Mean Square	F	Sig.
1	Regression	893.102	1	893.102	8.123	.006(a)
	Residual	5277.618	48	109.950		
	Total	6170.720	49			

a Predictors: (Constant), HE3-Meaningful participation in home
b Dependent Variable: Alienation
c Types of School = Convent School

4.3B: Coefficients

Model		Unstandardized Coefficients		Standardized Coefficients	t	Sig.
		B	Std. Error	Beta		
1	(Constant)	70.908	7.884		8.994	.000
	HE3-Meaningful participation in home	-2.596	.911	-.380	-2.850	.006

a Dependent Variable: Alienation

b Types of School = Convent School

Table–4.3B of coefficients shows the beta value of predictor variable i.e., β= -.380, it means that significant predictor variable, namely, 'Meaningful participation in home' was found negatively related with alienation. This finding seems to be highly logical as higher the students' meaningful participation in home lesser the alienation was found within them and lesser the meaningful participation in home higher the alienation may be generally experienced by any one, especially in the present case of students of Convent school.

In order to obtain more in-depth understanding with regard to alienation, school environment, home environment and their dimensions public and Convent school students and male and female students were compared on their mean scores to find significant differences using t-test.

Table–4.4 explains the mean differences between public and Convent school students on alienation, school environment, home environment and their dimensions.

Table- 4.4: Showing Differences between the mean scores of Public and Convent school students on Alienation, School environment, Home environment and their dimensions

Variables	Types of School	N	Mean	Std. Deviation	t	Sig
Alienation	Public School	50	45.70	9.355	-1.520	.132
	Convent School	50	48.84	11.222		
School Environment	Public School	50	23.12	3.734	-.514	.608
	Convent School	50	23.44	2.331		
SE1-Caring	Public School	50	7.66	1.965	-5.278**	.000
relationship in school	Convent School	50	9.56	1.618		
SE2- Expectation	Public School	50	8.16	2.122	.826	.411
in school	Convent School	50	7.84	1.730		
SE3-Meaningful	Public School	50	7.30	2.501	3.173**	.002
participation in school	Convent School	50	6.04	1.277		
Home Environment	Public School	50	20.00	3.505	-7.866**	.000
	Convent School	50	24.76	2.454		
HE1-Caring	Public School	50	7.20	1.750	-2.228*	.028
relationship in home	Convent School	50	8.00	1.841		
HE2-Expectaion	Public School	50	6.02	1.755	-6.868**	.000
in home	Convent School	50	8.26	1.496		
HE3-Meaningful	Public School	50	6.78	1.765	-5.042**	.000
participation in home	Convent School	50	8.50	1.644		

**P<.01, *P<.05

From the above table, it can be seen that public school students differ significantly with Convent school students on 'caring relationship in school', 'meaningful participation in school' - dimensions of school environment, 'home environment as a whole' and it's all three dimensions, namely, 'caring relationship in home', 'expectation in home', 'and meaningful participation in home'. It is clear from their mean values that public school students scored low in 'home environment as a whole' and their all three dimensions, namely, 'caring relationship in home', 'expectation in home', 'and meaningful participation in home' as well as 'caring relationship in school' compare to Convent school students whereas Convent school students scored low in 'meaningful participation in school' (Table- 4.4). On the 'total home environment, mean scores of public school students was low (M=20.00) when compared to mean scores of Convent school students (M=24.76) therefore, we can infer from their mean scores that Convent school students having more favorable home environment as compared to public school students. Similarly, we can also say that Convent school students' mean scores were high on 'caring relationship in school', 'caring relationship in home', 'expectation in home', 'and meaningful participation in home' as compared to public school students. Whereas public school students having high mean scores (M=7.30) on 'meaningful participation in school' as compared to Convent school students (M=6.04). Furthermore, Table-4.4 also clearly indicates that both school students (Public and Convent) did not differ significantly with each other on 'Alienation', 'School environment as a whole' and 'Expectation in school'

After explaining the significant differences between Public and convent school students, we described the differences between male and female students. Table–4.5 shows the mean differences between male and female students on alienation, school environment, home environment and their dimensions.

From the Table-4.5, it can be clearly observed that male students differ significantly from female students on 'caring relationship in school' – a dimension of school environment, 'home environment as a whole' and it's two dimensions, namely, 'expectation in home', 'and meaningful participation in home'. It is clear from the mean values that male students having low mean scores on 'caring relationship in school' 'home environment as a whole', 'expectation in home', 'and meaningful participation in home' (M = 8.27, M=21.48, M=6.69, M = 7.30) respectively, as compared to female students. The mean scores of female students on these variables are differ M = 9.30, M = 24.21, M=8.06, M = 8.33 respectively. Moreover, both groups did not differ significantly different on 'Alienation', 'School environment', 'Expectation in school', 'Meaningful participation in school' and 'Caring relationship in home'.

Table- 4.5: Showing differences between the mean scores of male and female students on Alienation, School environment, Home environment and their dimensions

Variables	Gender	N	Mean	Std. Deviation	t	Sig.
Alienation	Male	67	47.00	10.632	-.368	.713
	Female	33	47.82	10.048		
School Environment	Male	67	23.16	3.292	-.530	.597
	Female	33	23.52	2.706		
SE1-Caring relationship in school	Male	67	8.27	1.888	-2.457*	.016
	Female	33	9.30	2.158		
SE2- Expectation in school	Male	67	8.04	2.041	.329	.743
	Female	33	7.91	1.721		
SE3-Meaningful participation in school	Male	67	6.85	2.337	1.245	.216
	Female	33	6.30	1.357		
Home Environment	Male	67	21.48	3.894	-3.533**	.001
	Female	33	24.21	3.049		
HE1-Caring relationship in home	Male	67	7.49	1.910	-.835	.406
	Female	33	7.82	1.667		
HE2-Expectaion in home	Male	67	6.69	2.024	-3.447**	.001
	Female	33	8.06	1.519		
HE3-Meaningful participation in home	Male	67	7.30	1.801	-2.630**	.010
	Female	33	8.33	1.947		

**P<.01,   *P<.05

Chapter 5
SUMMARY, CONCLUSIONS AND FUTURE RESEARCH SUGGESTIONS

5.1 Summary

The present study is aimed at finding out *"Alienation of students in Public and Convent school students in relation to School Environment and Home Environment"*. Adolescence can be a time of both disorientation and discovery. Adolescence marks the transition between childhood and adulthood. By its very nature, it involves many physiological, psychological, social, and cognitive changes. These changes include the formation of a personal identity, the establishment of new peer networks, and the development of abstract thinking skills (Dacey & Kenny, 1997; Geldard & Geldard, 1999; Heaven, 1994). To manage these challenges, adolescents rely on their coping repertoire, which includes their problem solving competencies and skills.

Adolescence is a stage in the life span through which individuals pass in their preparation for adulthood. It is especially dynamic period because many of the roles of adolescents learning are unique to this time of life. The capacity to achieve a high level of self awareness arises during adolescence and self examination is the key to attaining maturity. Energies must be marshaled carefully. However, because consequences are cumulative, options are finite and every choice reduces later freedom.

A comprehensive understanding of adolescence also requires knowledge of youth's consciousness or the personal factors, aspirations, attitudes, beliefs, dispositions that enable young people to sustain intimate and lasting social relationship, accept vocational and economic responsibilities and form ideological convictions.

All young people have basic needs that are critical to survival and healthy development. They include a sense of safety and structure, belonging and membership; self worth and an ability to contribute; independence and control over one's life; closeness and several good relationships and competency and mastery. At the same time to succeed as adults all youth must acquire positive attitudes and appropriate behavior and skills in five areas: health, personal/social knowledge, reasoning and creative, vocation and citizenship. (Politz, 1996).

While all adolescents experience potentially alienating contexts, there appears to be a threshold at which adolescents are no longer able to tolerate what is happening to them, and they exhibit behaviors commonly associated with alienation (Tripp, 1986).

The present study is aimed at finding out the "Alienation of students in Public and Convent school students in relation to School Environment and Home Environment". The variables like alienation, home environment and school environment need to be describing briefly

Alienation

Alienation has been used in a variety of ways by philosophers, theologian, sociologist and psychologists. The alienated person as described by Fromm (1963), does not experience himself as the centre of his world, as the creator of his own acts but his acts and their consequences have become his masters, whom he obeys, or whom he even worship. For Seeman (1959) alienation has five components, namely, powerlessness, meaninglessness, normlessness, isolation and self estrangement, later Seeman (1972) revised alienation in six categories: (a) powerlessness – the sense of low control Vs. mastery over events, (b) meaninglessness the sense of incomprehensibility Vs. understandings of personal and social affairs,

(c) normlessness high expectancies for (or commitment to) socially unapproved means Vs. conventional means for the achievement of given goals, (d) cultural estrangement called value isolation the individual's rejection of commonly held value is the society Vs. commitment, (e) self estrangement the individual's engagement in activities that are not intrinsically rewarding Vs. involvement in a task of activity for its own sake, and (f) social isolation – the sense of exclusion or rejection Vs. social acceptance (Seeman, 1975).

Alienation is a term used to describe student estrangement in the learning process (Brown, Higgins, & Paulsen, 2003). Mann (2001) defined alienation as "the state or experience of being isolated from a group or an activity to which one should belong or in which one should be involved". The alienated person as described by Fromm (1963), does not experience himself as the centre of his world, as the creator of his own acts but his acts and their consequences have become his masters, whom he obeys, or whom he even worship. Alienation can affect adolescents in their home, social life, and future opportunities however, this study was most concerned with non-completers' experiences of alienation from a high school setting (Bronfenbrenner, 1986).

Langman and Kalekin-Fishman (2006) viewed social alienation as rooted in dysfunctional communication and power within a civil society. Alienation was returned to an individual need and social response for recognition. While the dimensions of alienation varied depending upon age, socio-economic status, race, and gender, an underlying commonality existed, the relationship between the individual and society.

Alienated students are described as those individuals who live on the borders of the school (Tyree Smith & Goc-Karp, 1994), the student who considers him or herself an outcast socially (Rodriguez, 1997) and academically, or "students who are reluctant to maintain the charade of acceptable behavior in school" (Carley, 1994). An alienating school environment can produce negative and disruptive behaviors which may lead to non-completion. These behaviors include hostility, passivity, withdrawal, lack of sense of responsibility for learning and getting work done, poor quality work, disinterest, and lack of involvement and initiative.

Viewing non-completion as a possible resultant behavior of school alienation may explain the high number of underrepresented youth who express feelings of non-attachment and lack of relevancy in public high school (Kawakami, 1994).

School Environment

Adolescents spend a large proportion of their day in school or pursuing school-related activities. While the primary purpose of school is the academic development of students, its effects on adolescents are far broader, also encompassing their physical and mental health, safety, civic engagement, and social development. Further, its effects on all these outcomes are produced through a variety of activities including formal pedagogy, after-school programs, caretaking activities (e.g., feeding, providing a safe environment) as well as the informal social environment created by students and staff on a daily basis. While most reports focus on a particular aspect of the school environment (e.g., academics, safety, health promotion), this brief looks at schools more comprehensively as an environment affecting multiple aspects of adolescent development.

A safe environment is a prerequisite for effective learning, so much so that the country's major education reform initiative, No Child Left Behind, requires school systems to have programs in place to reduce levels of violence as part of its larger plan to improve academic performance (U.S. Department of Education, 2007).

Feeling alienated from the school environment is a negative consequence of these factors or combination of factors which exists in schools (Calabrese & Noboa, 1995; Staples, 2000; Wehlage & Rutter, 1986) and will continue to "negatively affect the self-confidence, self-esteem, and self-direction of some students" (Brown et al., 2003). Behaviors which may indicate a student might feel alienated from the system are numerous and varied.

When students find their school environment to be supportive and caring, they are less likely to become involved in substance abuse, violence, and other problem behaviors (Hawkins, Catalano, Kosterman, Abbott, & Hill 1999; Battistich & Hom 1997; Resnick et al. 1997). They are more likely to develop positive attitudes toward themselves and prosocial attitudes and behaviors toward others (Schaps, Battistich, & Solomon 1997). Much of the available research shows that supportive schools foster these positive outcomes by promoting students' sense of "connectedness"

(Resnick et al. 1997), "belongingness" (Baumeister & Leary 1995), or "community" (Schaps, Battistich, & Solomon 1997) during the school day.

The literature supports the idea that the school environment and school processes impact the degree of student connectedness and alienation. And while much attention has been paid to the study of students' intellectual abilities and social development, "less attention has been paid to how well schools engage students in school life and how this affects their outlook on schooling and the future" (Thomson, 2005)

Home Environment

The role of home environment in fostering positive academic beliefs (values, competence perceptions, and goals) and behaviors is seen as important to keeping students engaged in school during adolescence. Research has shown that parents' beliefs and values can have an important impact of their children's achievement-related beliefs and values (Ames & Archer, 1987; Parson, Adler & Kaczala, 1982). In addition, direct parent involvement in and support of the child's school experiences are known to play an important role in a child's academic success (Epstein, 1989).

When a child stays at home all day, and parents take on the dual role of teacher and parent, issues of discipline will arise. It is easy for the child to take on a negative attitude towards understanding discipline. Correct discipline needs to be adhered from the start of homeschooling, to avoid potential difficulties later on in the child's development and learning.

Discipline provides both parents and child with immense levels of freedom, and there will be an enticement to stretch this freedom. Certain rules and practices need to be implemented at the beginning stages of the child as it may be very difficult for parent to change their child's habits at later stage.

There should be a friendly and enabling environment at home. The members of the family should listen and have a great respect for one another. This is very important to maintain a good environment free from all misunderstandings and confusions among the family members. Parents' relations play a significant role in maintaining a better environment at home. Negative relations and disputes can spoil the atmosphere of a house. Instead of a better environment the entire house plunges into chaos which not only affects the psychology of children but also makes them suffer in various complexes.

Research Objectives

The present research is systematically designed in accordance with the following main research objectives:

1. To identify alienation among public school students.

2. To identify alienation among convent school students.

3. To examine the difference between mean scores of Public and Convent School students on alienation.

4. To examine the difference between mean scores of Public and Convent School students on School Environment.

5. To examine the difference between mean scores of Public and Convent School students on Home Environment.

6. To identify the influences of public school and convent school students on alienated adolescents.

Research Questions

1. Does alienation exist in Public school student?

2. Does alienation exist in Convent school student?

3. Does school environment influence Public School alienated students?

4. Does school environment influence Convent School alienated students?

5. Does home environment influence Public School alienated students?

6. Does home environment influence Convent School alienated students?

7. Is there any difference exist between Public and Convent school students on alienation?

8. Is there any difference exist between Public and Convent school students on School Environment?

9. Is there any difference exist between Public and Convent school students on Home Environment?

Operational Definitions

Alienation

Alienation is "the state or experience of being isolated from a group or an activity to which one should belong or in which one should be involved" (Mann, 2001).

School Environment

School environment is an external protective factor, which in the present research, is defined as an environment, where an adolescents student experiences caring relationships with and health expectations from the school faculty and takes meaningful participation in the school and class related matters.

Home Environment

Home environment is defined as an environment where an adolescent member experiences caring relationships with and healthy expectations from the family members and indulge in meaningful participation in family related matters

Methodology

Subject

The data sample for this study consists of 300 adult students. Subjects were drawn from Al-Barkaat Public School and convent Our Lady of Fatima School of Aligarh (UP), India. Subjects were classified on the basis of scores obtained on alienation scale on which Quartile 1 and Quartile 3 were calculated to get the alienated subjects. Public school subjects whose scores fell below Q1 (45) were identified as low alienated and subjects whose scores fell above Q3 (52) were categorized as high alienated. Convent school subjects who have obtained scores below 41 were categorized as low alienated and those subjects whose score were above 49 they categorized as high alienated subjects. The sample further classified into two categories i.e. 50 students were drawn from convent schools and other 50 were taken from public schools.

The following tools were used in the present study

Alienation Scale

Alienation scale was developed by Kureshi and Dutt (1979).This scale comprised 21 items with four alternative response categories (always, often, sometimes, never, representing the five factors despair, disillusionment, unstructural universe, psychological vacuum and narcissum.

Home Environment and School Environment Scales

The Home and School Environment questionnaires are subscales of the Resilience and Youth Development Module (RYDM) which is a component of California Healthy Kids Survey (WestEd, 2002). The RYDM is devoted completely to assessing the internal and external assets associated with positive youth development and resilience (WestEd, 2002). The RYDM provides comprehensive and balanced coverage of external assets in home and school environment.

The full RYDM contains 59 questions that measure 17 external and 6 internal assets in the home, school, community, peer group and in the individual domains. Both, the home and school environment scales for measuring external assets, have 9 items each, and measure three common dimensions: *Caring relationships, High expectations and Meaningful participation.* Each item has four response options (very much true, pretty much true, a little true and not at all true) out of which the participants had to choose the one option which they felt best applied to them.

The investigator established rapport with the respondent students, prior to administer the proposed scales, and assured them that their responses would be kept strictly confidential and would be utilized for the research purpose only. After establishing rapport with the students, the data were collected in their classrooms in many sessions.

5.2. Conclusions

Findings of the present study based on statistical analysis with regards to the dependent variable, namely, alienation, and independent variables, namely, school environment and home environment have led to certain conclusion:

- ◆ 'Expectation in home'- as a dimension of home environment was found significant predictor of alienation among students.

- ◆ Significant predictive influences of predictor variables, namely, 'Caring relationship in school'- as a dimension of School environment and 'Home environment as a whole' on alienation of the students of public school were found statistically significant. 'Caring relationship in school' was negatively related to alienation whereas, 'Total home environment was found to be positively related to alienation in public school students.

♦ 'Meaningful participation in home'- as a dimension of home environment emerged as significant predictor of alienation. 'Meaningful participation in home' was found negatively related with alienation in Convent school students.

♦ Public school students differ significantly from Convent school students on 'caring relationship in school', 'meaningful participation in school' –as the dimensions of school environment, 'Home environment as a whole' and it's all three dimensions namely, 'caring relationship in home', 'expectation in home', 'and meaningful participation in home'.

♦ Students representing to Public and Convent schools did not differ significantly other on 'Alienation', and to the 'School environment as a whole' and 'Expectation in school'.

♦ Male and female students differed significantly on 'caring relationship in school' – as a dimension of school environment. Further, they differed in relation to 'home environment as a whole' and its two dimensions namely, 'expectation in home', 'and meaningful participation in home'.

♦ Male and female students did not differ significantly on 'Alienation', 'School environment', 'Expectation in school', 'Meaningful participation in school' and 'Caring relationship in home'.

5.3. Future Research Suggestions

♦ Research studies on alienation in relation to home and school environments are rather scanty; this area needs further researches and specifically researches in the field of Applied Psychology.

♦ Studies should also be explored the psychosocial causes of alienation across the life span.

♦ A number of personality variables such as anxiety, depression and hopelessness may be linked to alienation among adolescents.

♦ Influences of home and school environments on creativity, problem solving and academic achievement should be studied among adolescents.

♦ Future research studies should examine the influence of sociodemographic variables such as type of family, gender, locale, type of school, socioeconomic status etc. among adolescents' feelings of alienation.

REFERENCES

Abdallah, T. (1997). Reliability and validity of Palestinian Student Alienation Scale. *Adolescence, 32,* 367-371.

Adams, J. H. & Hannum, E. C., (2007). "Girls in Gansu, China." 71-98 in Exclusion, Gender and Education: Case Studies from the Developing World.

Adams, P. T. (1992). Public school/home school partnership: A study of the home school community in Maricopa County (Arizona). Unpublished EdD, Arizona State University.

Alston, J. A. (2004). The many faces of American schooling: Effective schools   research and border-crossing in the 21st century. *American Secondary   Education, 32(2),* 79-93.

American School Counselor Association [ASCA]. (2003). Taking your schools temperature: How school climate affects students and staff. Alexandria, VA: Author.

Ames , C. & Archer, J. (1987). Mother's beliefs about the role of ability and effort in school learning. *Journal of Educational Psychology, 79,* 409-414.

Ancess, J. (2003). Beating the Odds: High Schools as Communities of Commitment.   New York: Teachers College Press.

Anderson, P. L., & Cotton, C. S., (2001). Failing schools in Michigan: The surprising scale. Anderson Economic Group: Lansing, Michigan. 40.

Arnett, J. (1996). From the mouths of the metalheads: Heavy metal music and adolescent alienation. Boulder, CO: Westview Press.

Ascher, C. (1982). Student alienation, student behavior and urban schools. ERIC Clearinghouse on Urban Education, Box 40, Teachers College, Columbia University, New York, NY. (ERIC Document Reproduction Service No. ED 219 484).

Bahr, S., Hawks, R., & Wang, G. (1993). Family and religious influences on adolescent substance abuse. *Youth and Society, 24*, 443-465.

Barratt-Peacock, J. (1997). The why and how of Australian home education. Yankalilla, SA: Learning Books.

Battistich, V., & Hom, A. (1997). The relationship between students' sense of their school as a community and their involvement in problem behaviors. *American Journal of Public Health, 87(12)*, 1997–2001.

Baumeister, R., & Leary, M. (1995). The need to belong: Desire for interpersonal attachments as a fundamental human motivation. *Psychological Bulletin, 117(3)*, 497–529.

Blackledge, A. (2001). "The Wrong Sort of Capital." International Journal of Bilingualism, 513, 345-69.

Booker, K. C. (2007). Likeness, comfort, and tolerance: Examining African American adolescents' sense of school belonging. Urban Review: Issues and Ideas in Public *Education, 39(3)*, 301-317.

Bronfenbrenner, U. (1974). The ecology of human development. Cambridge, Massachusetts: Harvard University Press.

Bronfenbrenner, U. (1986). Alienation and the four worlds of childhood. Phi Delta Kappan, 67, 430-436.

Brosna, P. (1991). Child competencies and family processes in home school families. Unpublished MEd, University of Melbourne, Melbourne.

Brown, M. R., Higgins, K. & Paulsen, K. (2003). Adolescent alienation: What is it and what can educators do about it? *Intervention in School and Clinic, 39(1)*, 3-9.

Bulach, C., Malone, B., & Castleman, C. (1995). An investigation of variables related to student achievement. *Mid-Western Educational Researcher, 8(2)*, 23 29.

Calabrese, R. L. (1990). Adolescence: A growth period conducive to alienation. *Adolescence, 22*, 929-938.

Calabrese, R. L., & Adams, J. (1990). Alienation: A cause of juvenile delinquency. *Adolescence, 25*, 435-440.

Calabrese, R. L., & Noboa, J. (1995). The choice for gang membership by Mexican-American adolescents. *High School Journal, 78(4)*, 226-235.

Calabrese, R. L., & Seldin, C. A. (1987b). A contextual analysis of alienation among school constituencies. *Urban Education, 22*, 227-237.

Carley, G. (1994). Shifting alienated student-authority relationships in a high school. *Social Work in Education, 16(4)*, 221-230.

Carlson, T. B. (1995). We hate gym: Student alienation from physical education. *Journal of Teaching in Physical Education, 14(4)*, 467-477.

Carnegie Council on Adolescent Development. (1989). Turning points: Preparing American youth for the 21st century. New York: Carnegie Corporation.

Carter, P. (2003). Black Cultural Capital, Status Positioning, and Schooling Conflict Conflicts for Low-Income African American Youth. *Social Problems 5011(2003)*, 136-55.

Cassidy, T., & Lynn, R. (1991). Achievement motivation, educational attainment, cycles of disadvantage and social competence: Some longitudinal data. *British Journal of Educational Psychology, 61*, 1-12.

Center for Collaborative Education. (2003). How are Boston Pilot Schools Students Faring?

Chaskin, R. J., & Rauner, D. M. (1995). Youth and caring. *Phi Delta Kappan, 76(9)*, 667-675.

Clery, E. (1998). Homeschooling: The meaning that the homeschooled child assigns to this experience. *Issues in Educational Research, 8(1)*, 1-13.

Coburn, J. & Nelson, S. (1989). Teachers do make a difference: What Indian graduates say about their school experience. Portland: Northwest Regional Educational Laboratory. (ERIC Document Reproduction Service No. ED 306 071).

Coley, R. J. (1995). Dreams deferred: High school dropouts in the United States. Princeton, NJ: Educational Testing Service Policy Information Center.

Comer, J. (1998). Educating poor minority children. *Scientific American, 259*, 42-48.

Constantine, M. G., Kindaichi, M. M., & Miville, M. L. (2007). Factors influencing the educational and vocational transitions of Black and Latino high school students. *Professional School Counseling, 10*, 261-265.

Constantine, N. A., & Benard, B. (2001). California Healthy Kids Survey Resilience Assessment Module: Technical report. Berkeley, CA: Public Health Institute

Cothran, D., & Ennis, C. (2000). Building bridges to student engagement: Communication respect and care for students in urban high schools. *Journal of Research and Development in Education, 33, (2),* 106-117.

Cotton, K. 1996. School Improvement Research Series (SIRS) School Size, School Climate, and Student Performance.

Crinson, I., & Yuill, C. (2008). What can alienation theory contribute to an understanding of social inequalities in health? *International Journal of Health Services, 38(3),* 455-469.

Cunnigham, W. C., & Cordeiro, P. A. (2000). Educational administration: A problem-based approach. Boston: Allyn and Bacon.

Dacey, J., & Kenny, M. (1997). Adolescent development. Chicago: Brown & Benchmark.

Darling-Hammond, L. (1997). The Right to Learn: A Blueprint for School Reform. San Francisco, Jossey-Bass.

Darling-Hammond, L. (2002). Redesigning Schools: What Matters Most and What Works School Redesign Network.

Davis, S. J. H. (2000). A survey of public school educators of the needs of home-schooled students entering the public school at the secondary level. Unpublished EdD, Wilmington College, Delaware.

Davison Aviles, R. M., Guerrero, M. P., Howarth, H. B., & Thomas, G. (1999). Perceptions of Chicano/Latino students who have dropped out of school. *Journal of Counseling & Development, 77(4)*, 465-473.

De Block, A & Adriaens, P. P. (2011). Maladapting minds: Philosophy, psychiatry, and evolutionary theory. New York, NY: Oxford University Press.

Dean, D. G. (1961). Alienation: Its meaning and measurement. *American Sociological Review, 25*, 753-758.

Downey, D. & Pribesh, S. (2004). When Race Matters: Teachers' Evaluations of Students' Classroom Behavior. *Sociology of Education, 77(4)*, 267-82.

Eccles, J. S., Midgley, C., Wigfield, A., Buchanan, C.M., Reuman, D., Flanagan, C. & Maclver, D., (1993). Development during adolescence: The impact of stage-environment fit on adolescents' experiences in schools and families. *American Psychologist, 48*, 90-101.

Eccles, J.S., Wigfield, A., Schiefele, U., (1998). Motivation to succeed. In W. Damon (Series Ed.) & N. Eisenberg (Vol. Ed.) Handbook of child psychology: Vol. 3. Social, emotional and personality development (5th ed., 1017-1095). New York: Wiley.

Education Queensland, (2003). Home Schooling Review (Research-Parliamentary report). Brisbane.

Education Trust. (2002). Achievement in America. Washington, DC: Author.

Edwards, D., & Mullis, F. (2001). Creating a sense of belonging to build safe schools. *Journal of Individual Psychology, 57(2)*, 196-203.

Englund, M., Luckner, A., Whaley, G., and Egeland, B. (2004). Children's Achievement inEarly Elementary School: Longitudinal Effects of Parental Involvement, Expectations, and Quality of Assistance. *Journal of Educational Psychology, 94 (4)*, 723-730.

Entwisle, D. R., Karl, L. Alexander, & Olson, L. S. (1997). *Children, Schools, and Inequality.* Westview Press.

Epstein, J. L. (1983). Longitudinal effects of family-school-person interactions on student outcomes. *Research in Sociology of Education and Socialization, 4*, 101-127. Greenwich, CT: JAI Press.

Epstein, K. K. (1992). Case studies in dropping out and dropping back in. *Journal of Education, 174(3)*, 55-65.

Esptein , J. L. (1989). *Family influence and student motivation*. In R. E. Ames & C. Ames (Eds.), Research on motivation in education (3) NY: Academic Press.

Fairchild, E. E. (2002). Home schooling and public education in Iowa: The views of rural superintendents. Unpublished PhD, The University of Iowa.

Ferguson, R. (1991). Paying for public education: New evidence of how and why money matters. *Harvard Journal on Legislation, 28*, 465-98.

Fetco, J.V. (1985). *Adolescent alienation: Assessment and application.* Paper presented at the annual meeting of the American Public Health Association, Washington, DC. (ERIC Document Reproduction Service No. ED 270 657)

Freiberg, H. J. (1998). Measuring school climate: Let me count the ways. *Educational Leadership, 56(1)*, 22-26.

Fromm, E. (1963). *Disobedient as a psychological and moral problem* (Ed.), Writing and reading Across the Curriculum. 357-361. New York: Pearson.

Fuller, B. & Clarke, P. (1994). Raising school effect while ignoring culture? Local conditions and the influence of classroom tools, rules, and pedagogy. *Review of Educational Research, 64*, 119-57.

Fuller, B. (1987). What School Factors Raise Achievement in the Third World? Review of Educational Research, 57, 255-92.

Geierstanger, S., Peterson, S., Amaral, M., Gorette, G., Mansour, Mona, W., & Russell, S. (2004). School-Based Health Centers and Academic Performance: Research, Challenges, and Recommendations. *Journal of School Health, 74(9)*, 347-352.

Geldard, K. & Geldard, D. (1999). *Counseling Adolescents*. London: Sage Publications.

Geray, D. W. (1998). A study of the academic achievements of home-schooled students who have matriculated into postsecondary institutions. Unpublished EdD, University of Sarasota.

Goddard, R., Hoy, W., & Hoy, A. (2000). Collective teacher efficacy: Its meaning, measure, and impact on student achievement. *American Educational Research Journal, 37(2),* 479-507.

Goodwin, B. (2000). *Raising the achievement of low-performing students (1-10)* Office of Educational Research and Improvement.

Gottfried, A. E., Fleming, J. S., & Gottfried, A. W. (1998). Role of parental motivational practices in children's aca- demic intrinsic motivation and achievement. *Journal of Educational Psychology, 86,* 104-113.

Goymer, S. (2001). *The legacy of home-schooling: Case studies of late adolescents in transition.* Unpublished PhD, East Anglia.

Greene, J., & Winters, M. (2005). Public High School Graduation and College- Readiness Rate: 1991-2002. New York: The Manhattan Institute.

Gullotta, T. P. (2001). Early adolescence, alienation, and education. *Theory into Practice, 22(2),* 151-154.

Habibullah, A. (2004). Mum, when's recess? A glimpse into two contexts of home schooling. Unpublished Honours of BEd, Monash, Melbourne.

Hallinan, M. T. (2008). Teacher influences on students' attachment to school. *Sociology of Education 81(7),* 271-83.

Hammer, B. (2003). ETS identifies affecting student achievement-Washington update.

Hanna, L. G. (1996). Home schooling in the 1990s: An issue of access. Unpublished EdD, Immaculata College.

Hansen, J. B., & Toso, S. J. (2007). Gifted dropouts: Personality, family, social, and school factors. *Gifted Child Today, 30(4),* 30-41.

Hanson, T. L., & Kim, J. O. (2007). Measuring resilience and youth development: The psychometric properties of the Healthy Kids Survey. (Issues & Answers Report, REL 2007–No. 034). Washington, DC: U.S. Department of Education, Institute of Education Sciences, National Center for Education Evaluation and Regional Assistance, Regional Educational Laboratory West.

Hanushek, E. (1995). Interpreting recent research on schooling in developing countries. *World Bank Research Observer, 10(2),* 247-54.

Harding, T. (2003). *A comparison of the academic results of students monitored by the State, with the academic results of students not monitored by the State.* In A submission for the Home Schooling Review, (6). Brisbane: Australian Christian Academy.

Harding, T. J. A. (1997). *Why Australian Christian Academy families in Queensland choose to home school: Implications for policy development.* Unpublished MEd, Partial fulfillment, University of Technology, Brisbane.

Harding, T., & Farrell, A. (2002). *Alternate models of schooling: Legal and ethical considerations.* Paper presented at the Australia and New Zealand Education Law Association. Conference (11th : 2002 : Brisbane Qld), Brisbane Qld.

Harp, B. (1998). *Home schooling: A study of reasons why some central Queensland parents choose the home schooling alternative for their children.* Unpublished Master of Education Studies, part fulfillment, Central Queensland University, Rockhampton.

Hauser-Cram, P., Sirin, S. & Stipek, D. (2003). *"When Teachers' and Parents'*

Hawkins, J. D., Catalano, R. F., Kosterman, R., Abbott, R., & Hill, K. G. (1999). Preventing adolescent health-risk behaviors by strengthening protection during childhood. *Archives of Pediatric and Adolescent Medicine, 153,* 226–234.

Hawkins, J. D., Herrenkohl, T. I., Farrington, D. P., Brewer, D., Catalano, R. F., Harachi, T. W., & Cothern, L. (2000). Predictors of youth violence. *Juvenile Justice Bulletin,* ERIC ED440196. Washington, DC: Department of Justice.

Hawkins, J. D., Herrenkohl, T., Farrington, D. P., Brewer, D., Catalano, R. F., Harachi, T. W., & Cothern, L. (2000). *Predictors of violent youth.* (ED440196).

Heaven, P. C. (1994). *Contemporary adolescence: A social psychological approach.* Melbourne, Australia: Macmillan Education Australia.

Heck, R. (2000). Examining the impact of school quality on school outcomes and improvement: A value-added approach. *Educational Administration Quarterly, 36(4),* 513-552.

Hedegaard, M. (2005). *Child development from a cultural-historical approach: Children's activity in everyday local settings as foundation for their development.* Unpublished manuscript, Melbourne.

Hoffman, M.A., Levy, S. R., & Malinsky, D. (1996). Stress and adjustment in the transition to adolescents: Moderating effects C, neuroticism and extraversion. *Journal of Youth and Adolescence, 25(2),* 161-175.

Honeybone, R. (2000). *A South Australian case study examining the home - schooling experiences of eight primary school aged children and their families.* Unpublished Thesis B.Ed. (Hons.), University of South Australia, Adelaide.

Howse, R. B. (1999). *Motivation and self-regulation as predictors of achievement in economically disadvantaged young children.* Dissertation Abstract International, 60-06B, 2985.

Hoy, W., & Miskel, C. (2005). *Education administration: Theory, research, and practice* (7th ed.). New York: McGraw Hill.

Hoy, W., Tarter, C., & Bliss, F. (1990). School characteristics and faculty trust in secondary schools. *Educational Administration Quarterly, 25,* 294-309.

Hughes, J., Gleason, K., & Zhang, D. (2005). "Relationship Influences on Teachers' perceptions of Academic Competence in Academically at-risk Minority and Majority First Grade Students." *Journal of School Psychology, 43,* 303-20.

Israel, J. (1972). Alienation from Marx to modern sociology. Boston: Allyn and Bacon.

Jacob, A., Barratt-Peacock, J., Carins, K., Holderness-Roddam, G., Home, A., & Shipway, K. (1991). Home education in Tasmania: Report of ministerial working party October 1991. Hobart: Government Printer.

Jeffrey, D., Giskes, R. & Section (2004). *Home schooling. Queensland Parliamentary Library, Research Publications and Resources Section*

Jerald, C. D. (2000). State of the states. *Education Week, 19(18),* 62-65.

Jessor, R., & Jessor, S. L. (1977). *Problem behavior and psychosocial development.* New York: Academic Press.

Johnson, L. S. (2009). School contexts and student belonging: A mixed methods study of an innovative high school. *The School Community Journal, 19(1)*, 99-118.

Johnston, Janet R. (2005). Children of Divorce Who Reject a Parent and Refuse Visitation: RecentResearch and Social Policy Implications for the Alienated Child. *Family Law Quarterly, 38,757-775.*

Jordan, W. J., Lara, L. & James, M. & McPartland. (1996). Exploring the causes of early dropout among race-ethnic and gender groups. *Youth and Society 28*, 62–94.

Kelley, R. C., Thorton, B., & Daugherty, R. (2005). Relationships between measures of leadership and school climate. *Education, 126(1)*, 17-25.

Kingston, P. (2001). The Unfulfilled Promise of Cultural Capital Theory. *Sociology of Education (Extra Issue)(2001)*, 88–99.

Knesting, K. (2008). Students at risk for school dropout: Supporting their assistance. *Preventing School Failure, 52(4)*, 3-10.

Krivanek, R. (1988). *Social development in home based education.* Unpublished MA, University of Melbourne, Parkville Vic.

Krout, A. M. (2001). *Home scholars transition to public schools in West Virginia.* Unpublished EdD, West Virginia University.

Kunkel, R. C., Thompson, J. C., & McElhinney, J. H. (1973). *School related alienation: Perceptions of secondary school students,* Paper presented at the 57th annual meeting of the American Educational Research Association. New Orleans, LA.

Kureshi, A. & Dutt, M. (1979). *Dimensions of Alienation-A Factor-analytic study.* Psychologia, 99-105.

Lacourse, E., Villeneuve, M. & Claes, M. (2003), Department of Psychology, University of Montreal. Requests for reprints should be sent to Eric Lacourse, G.R.I.P. Universite de Montreal, Unite de recherche biopsychosociale, Hopital Ste-Justine, 3175, Chemin de la Cote Ste-Catherine, Montreal, Quebec, Canada H3T 1C5. E-mail: eric.lacourse@umontreal.ca COPYRIGHT 2003 Libra Publishers, Inc. COPYRIGHT 2004 Gale Group BIBLIOGRAPHY

Langman, L., & Kalekin-Fishman, D. (2006). *Introduction.* In R. Langman & D. Kalekin-Fishman (Eds.), The evolution of alienation: Trauma, promise, and the millennium (pp. 2-20). Lanham, MD: Rowman & Littlefield Publishers, Inc.

Lareau, A, & Weininger, W. (2003). Cultural Capital in Educational Research: A Critical Assessment. *Theory and Society, 32,* 567–606.

Lareau, A., & Horvat, E. M. (1999). Moments of Social Inclusion and Exclusion: Race, Class,. and Cultural Capital in Family-School Relationship. *Sociology of Education, 72,* 37–53.

Lashway, L. (2003). The mandate to help low-performing schools: ERIC Clearinghouse on Educational Management.

Lattibeaudiere, V. H. (2000). *An exploratory study of the transition and adjustment of former home-schooled students to college life.* Unpublished PhD, The University of Tennessee.

Leffert, N., Benson, P., L., & Roehlkepartain, J. L. (1997). *Starting out right: Developmental assets for children.* Minneapolis, MN: Search Institute.

Mackey, J., & Ahlgren, A. (1977). Dimensions of adolescent alienation. *Applied Psychological Measurement, 1,* 219-232.

Mann, S J .(2001). Alternative perspective on the student experience: Alienation and engagement. *Studies in Higher Education 26,*7-13

Margalit, M. (1991). Understanding loneliness among students with learning disabilities. *Behavioral Change, 8(4),* 167-173.

Mau, R. Y. (1992). The validity and devolution of a concept: Student alienation. *Adolescence, 27,* 721-741.

Mayer, D. & Ralph, J. (2007). *"Key Indicators of School Quality".* In "Key Indicators of Child and Youth Well-Being", edited by Brett V. Brown. Taylor and Francis, NY: 2007.

McColl, A. (2005). *ACE homeschooling:* The graduates speak. Unpublished Masters of Education, part fulfillment, Christian Heritage College, Brisbane.

McInerney, P. (2009).Towards a critical pedagogy of engagement for alienated youth: Insights from Freire and school-based research. *Critical Studies in Education, 50(1),* 23-35.

McMillan, J. H. (1992). A qualitative study of resilient at-risk students: Metropolitan Educational Research Consortium.

McMillan, J. H., & Reed, D. A. (1994). At-risk students and resiliency: Factors contributing to academic success. *The Clearing House, 67(3),* 137-140.

National Institute enter for Education Statistics. (2010). National assessment of educational progress. Washington, DC: U.S. Department of Education.

Niebuhr, K. (1995). *The effect of motivation on the relationship of school climate, family environment, and student characteristics to academic achievement.* (ERIC Document Reproduction Service ED 393 202).

Oerlemans, K., & Jenkins, H. (1998). There are aliens in our school. *Issues in Educational Research, 8(2),* 117-129.

Parson, J., Adler, T.F. & Kaczala, C.M. (1982). Socialization of achievement attitudes and beliefs: Parental Influences. *Child Development, 53,* 310-312.

Peng, S. S., & Wright, D. (1994). Explanation of academic achievement of Asian -American students. *Journal of Educational Research, 87 (6),* 346-352.

Phillips, M. (1998). *Family background, parenting practices, and the black-white test score gap. The black-white test score gab,* Washington, D.C., Brooking Institution Press.

Politz, B. (1996). Making the Case: Community Foundations and Youth Development, Academy for Educational Development, Centre for Youth Development & Policy Research, Foundations for Change, Second Edition.

Powell, A, Farrar, E, & Cohen, D.K. (1985). *The Shopping Mall High School: Winners and Losers in the Educational Marketplace.* Boston: Houghton Mifflin.

Price, H. B. (2008). *Mobilizing the community to help students succeed. Alexandria, VA:* Association for Supervision and Curriculum Development.psychometric properties of the Healthy Kids Survey. (Issues & Answers Report, REL

Rafalides, M. & Hoy, W. (1971). Student sense of alienation and pupil control orientation of high schools. *The High School Journal, 55,* 101-111.

Rayce, S.L.B., Holstein, B.E., & Kreiner, S. (2008). Aspects of alienation and symptom load among adolescents. *European Journal of Public Health, 19(1),* 79-84.

Reay, D. (2004). It's all becoming a habitus': Beyond the habitual use of habitus in educational research. *British Journal of Sociology of Education, 25(4),* 431-44.

Resnick, M. D., Bearman, P. S., Blum, R. W., Bauman, K. E., Harris, K. M., Jones, J., Tabor, J., Beuhring, T., Sieving, R. E., Shew, M., Ireland, M., Bearinger, L. H., & Udry, J. R. (1997). Protecting adolescents from harm: Findings from the National Longitudinal Study on Adolescent Health. *Journal of the American Medical Association, 278(10),* 823–832.

Roberts, B. R. (1987). A confirmatory factor analytic model of alienation. *Social Psychology Quarterly, 50,* 346-351.

Rodriguez, J. (1997). *At-risk: A measure of school failure in American education,* Annual Meeting of the American Educational Research Association. Chicago, IL.

Roeser, R. W. & Eccles, Jacquelynne, S. (1998). "Linking the study of schooling and mental health: Selected issues and empirical illustrations." *Educational Psychologist, 33(4),* 153.

Rogoff, B. (1990). *Apprenticeship in thinking: Cognitive development in social context.* New York: Oxford University Press. Isbn 0195070038, 9780195070033 length 242 pages

Rogoff, B. (2003). *The cultural nature of human development.* New York: Oxford University Press.

Rollins, B. C., & Thomas, D. L. (1979). *Parental support, power, and control techniques in the socialization of children.* In W. R. Burr, R. Hill, F. I. Nye, & I. L. Reiss (Eds.), Contemporary theories about the family, Vol. L (pp. 317-364). New York: The Free Press, Macmillan.

Schaps, E., Battistich, V., & Solomon, D. (1997). *School as a caring community: A key to character.* In A. Molnar (Ed.), The construction of children's character. Ninety-sixth yearbook of the National Society for the Study of Education (pp. 127–139). Chicago: National Society for the Study of Education.

Schiefele, U., Krapp, A., & Winteler, A. (1992). *Interest as a predictor of academic achievement: A meta-analysis of research.* The role of interest in learning and development (183-212). Hillsdale, NJ: Erlbaum.

Schunk, D. H., & Pajares, F. (2002). *The development of academic self-efficacy. Development of achievement motivation.* 15-32. San Diego, CA: Academic Press.

Scott, J., Terrance M., Nelson, C. Michael, & Liaupsin, Carl J. (2001). *Effective instruction: The forgotten component in preventing school violence.* Education & Treatment of Children, 24(3). Special issue: Severe behavior disorders of children and youth. 309- 322.

Seeman, M. (1959). On the meaning of alienation. *American Sociological Review, 24,* 783-791.

Seeman, M. (1972). *Alienation and engagement.* In .4 Campbell & P. E. Converse (Eds.), The human meaning of social change (467-527). New York: Russell Sage.

Seeman, M. (1975). Alienation studies. *Annual Review of Sociology, 1,* 91- 123.

Senechal, M., & LeFevre, J. (2002). *Storybook reading and parent teaching: Links to language and literacy devel-opment.* In P. R. Britto & J. Brooks-Gunn (Eds.), New directions in child development: No. 92. The role of family literacy environments in promoting young children's emerging literacy skills (pp. 39-52). San Francisco: Jossey- Bass.

Shannon, S. G., & Bylsma, P. (2002). *Addressing the achievement gap: A challenge for Washington state educators.* Olympia, WA: Office of the Superintendent of Public Instruction.

Shoho, A. R. (1996). *The alienation of rural middle school students: Implications for gang membership.* Paper Presented at the Annual Meeting of the American Educational Research Association (New York, NY, April 8-12). (ERIC Document Reproduction Service No. ED 396 889).

Sikkink, D. (1999). The social sources of alienation from public schools. *Social Forces, 78(1),* 51-86.

Simich, M. (1998). *How parents who home school their children manage the process.* Unpublished MEd, University-of-Western-Australia, Nedlands WA.

Sizer, R. T . (1984). *Horace's compromise: The dilemma of the American high school.* Boston: Houghton Mifflin.

Smith, T. B., & Goc.-Karp, G. (1994). *Becoming marginalized in a middle school physical education class,* Annual Meeting of the American Education research Association. New Orleans, Louisiana. 294.

Smokowski, P. R., & Kopasz, K. H. (2005). Bullying in school: An overview of types, effects, family characteristics, and intervention strategies. *Children and Schools, 27(2),* 101-110.

Smrekar, C. (1999). *School Choice in Urban America: Magnet Schools and the Pursuit of Equity.* Teachers College Press.

Snyder, J. W., III. (2005). *The perceived effectiveness of the transition process to public school education from home schooling.* Unpublished EdD, Widener University.

Stanton-Salazar, R, & Dornbusch, S., (1995). Social Capital and the Reproduction of Inequality: Information Networks Among Mexican-Origin High School Students. *Sociology of Education, 68,* 116-35.

Staples, S. J. (2000). Violence in schools: Rage against a broken world. *The Annals of the American Academy,* 31-41.

Stevenson, H. W., & Stigler, J. W., (1992). *The Learning Gap.* New York: Summit Books.

Stoppler, L. S. R. (1998). *Continuing and discontinuing home schooling: Meeting the needs of children.* Unpublished MEd, University of Northern British Columbia.

Sullivan, E. (2003). *Civil society and school accountability: A human rights approach to parent and community participation in nyc schools,* New York University Inst. for Education and Social Policy. New York University, NY.

Swaminathan, R. (2004). It's my place: Student perspectives on urban school effectiveness. *School Effectiveness and School Improvement, 15(1),* 33-63.

Taylor, E. D. (1999). *How does peer support relate to african american adolescent's academic outcomes?* Testing a conceptual model, Paper presented at the biennial meeting of the Society for Research in Child Development. Albuquerque, NM.

Thomas, A. (1998*). Educating children at home.* London: Cassell.

Thompson, G. L. (2007). *Up where we belong.* San Fransisco, CA: Jossey Bass.

Thomson, S. (2005). Engaging students with school life. *Youth Studies Australia, 24(1),* 10-15.

Tomaka, J., & Palacios, R. (2006). The relation of social isolation, loneliness, and social support to disease outcomes among the elderly. *Journal of Aging and Health,18(3),* 359-384.

Tripp, D. (1986). *Critical incidents in teaching.* London: Rutledge.

U.S. Department of Education. "No Child Left Behind: A Desktop Reference." Accessed September 28, 2007. Available at: <u>http://www.ed.gov/admins/lead/account/nclbreference/index.html</u>

U.S. Department of Education. National Center for Education Statistics. Monitoring School Quality: An Indicators Report, NCES 2001–030 by Daniel P. Mayer, John E. Mullens, and Mary T. Moore. John Ralph, Project Officer.Washington, DC: 2000.

Urban, V. (1999). A case for caring. *Educational Leadership, 56(6),* 69-70.

Valverde, S. A. (1987). A comparative study of hispanic high school dropouts and graduates: Why do some leave school early and some finish? *Education and Urban Society, 19(3),* 320-329.

Voisin, D. R., Salazar, L. F., Crosby, R., Diclemente, R. J., Yarber, W. L., & Staples-Horne, M. (2005). Teacher connectedness and health-related outcomes among detained adolescents. *Journal of Adolescent Health, 37,* 337.17-337-23.

Wang, J., Wildman, L., & Calhoun, G. (1996). The relationship between parental influences and student achievement in seventh grade mathematics. *School Science and Mathematics, 96 (8),* 395-400.

Wehlage, G . G., & Rutter, R. A. (1986). *Dropping out: How much d o schools contribute to the problem?* In Natriello, G. (Ed.), School dropouts: Patterns and policies, 70-88. New York: Teachers College Press.

Wehlage, G. G. (1986). At-risk students and the need for high school reform. *Education, 107(1),* 18-29.

Wenz, F. V. (1979). Sociological correlates of alienation among adolescent suicide attempts. *Adolescence, 14(53),* 19-30.

WestEd. (2002). *Resilience and Youth Development Module: Aggregated California data fall 1999 to spring 2002.* Los Alamitos, CA: Author. Retrieved June 1, 2006, from www.wested.org/chks/pdf/rydm_aggregate.pdf

Williamson, I., & Cullingford, C. (1997). The uses and misuses of "alienation" in the social sciences and education. *British Journal of Educational Studies, 45,* 263-275.

Williamson, I., & Cullingford, C. (1998). Adolescent alienation: Its correlates and consequences. *Educational Studies, 24,* 333—343.

Wilson, R. (1989). Ethnic diversity and differing value systems: A theoretical analysis of educational alienation. *Peabody Journal of Education, 66(4),* 42-50.

Young, T. J. (1985). Adolescent suicide: The clinical manifestation of alienation. *High School Journal, 69,* 55-59.

Zhang, Y., Kao, G., & Hannum, F. (2007). Do Mothers in Rural China Practice gender Equality in Educational Expectation for their Children? *Comparative Education Review, 51(2),* 131-58.

www.ingramcontent.com/pod-product-compliance
Lightning Source LLC
LaVergne TN
LVHW090857240726
843527LV00049B/68